T0283528

WHY WE LOVE

Also by Anna Machin
The Life of Dad: The Making of a Modern Father

WHY
WE
LOVE

The New Science Behind
Our Closest Relationships

DR. ANNA MACHIN

PEGASUS BOOKS
NEW YORK LONDON

WHY WE LOVE

Pegasus Books, Ltd.
148 West 37th Street, 13th Floor
New York, NY 10018

First Pegasus Books cloth edition February 2022

ISBN: 978-1-64313-922-7

10 9 8 7 6 5 4 3 2 1

Printed in the United States of America
Distributed by Simon & Schuster
www.pegasusbooks.com

For Hebe, Kitty and Lydia

In loving memory of my best boy, Bear.

CONTENTS

• • • •

PROLOGUE

• • • •

Love . . . it's complicated.

It's fair to say that this is not the first book written about love. Indeed, the shelves of bookshops and libraries are crammed with authors proffering their views on love from many differing perspectives; psychological, philosophical, scientific, cultural. During my years of studying love, I have read many of these books, and they have provided helpful insights and sent me down new routes of research. But what many of them have tried to do is provide *the* answer to the question 'What is love?' Love is regularly reduced to a set of chemicals in the brain, or an entirely cultural construct, or the route to great art and creativity. And this is unsurprising. We are a knowledge-hungry species who dislikes uncertainty. We are never happier than when we have a clear understanding of where we are going. But the thing about love is this: it's complicated.

As an anthropologist, my job is to observe my fellow humans and then explain as fully as I am able the cause of the behaviour or anatomical quirk I see in front of me. And this means I am a bit like a magpie, borrowing ideas and techniques from other human-focused disciplines to make sure I have sought out all the evidence that enables me to present an answer at all levels of explanation. The goal is 360° understanding. The result of this is that straightforward answers are often elusive. And the study of love is no different. All the disciplines of academia seem to have their own answer to the conundrum of love. But in contrast to other areas of study, where all these explanations can be a bit of a headache, when it comes to love my reaction is one of awe. I am in awe of the sheer immensity of love. In awe of the way it infiltrates every part of our life and every fibre of our being. In awe of how it sits at the very centre of our existence, such is its power to shape our health, happiness and life course. In awe of how we get to experience love in so many ways and with so many people, animals and beings. I think we are incredibly lucky.

So this book intends not to give you a single answer to the question 'What is love?' Instead of delivering a nice, neat explanation by reducing the cause to a single factor, it intends to do the exact opposite. This book gives you the expansionist answer. I want to present you with ten responses which separately give a strong and robustly evidenced answer to the questions which permeate our discussions about love. My aim is that by bringing these diverse answers together, and making it clear that no single one is the *complete* answer, I might just give you an inkling of the immensity and the true awesomeness of human love. All forms of love will be considered – romantic, platonic, spiritual, futuristic and parasocial – and all the scientific and social-scientific explanations interrogated. This does mean that at the end there will be no formula for love. No neat explanation that will guide your life and keep you on track and to timetable. But what I hope there will be is a reborn acknowledgement of the immensity of love, and a reconsideration of the many places where love exists in your life. Because I think we might have started taking love for granted, reducing it to a chore that we can efficiently tick off our list with the use of social media. And in the west our privileging of romantic love above all else has meant that maybe we have forgotten the other forms of love that we have in our lives – those with family, friends, pets, gods – which all go to making us who we are. Because that is part of the joy of being human. Unlike many of our fellow animals, we get to experience love in *so* many ways.

I will use evidence from across the disciplines to build my arguments, so the hard sciences of genetics, pharmacology and neuroscience will make regular appearances. We will also encounter psychology, philosophy, social anthropology and theology, because the explanation for any human behaviour or experience is inevitably multi-layered. So, yes, this is a science book, but more than that it is a book about a key aspect of the human condition. As a result, I hope there is something for everyone. And there is no need to be a scientist or an anthropologist to follow my arguments because we are *all* experts in love. To reinforce this, as well as giving you easy-to-follow summaries of what we academics know, you will also hear the voices of real people recounting their experiences of love and relationships with everyone from their child to their best friend, their dog to their god and even their favourite band. I hope you can add yours to them.

This book is about the why, how, what and who of love. It'll explain why love evolved in the first place and how all of our bodily mechanisms – behavioural, physiological, neural – are attuned to make sure we grab it and keep it. It'll unpack what causes love to be such a profoundly individual experience and explore the mechanisms – both biological and cultural – that make how I love and how you love so different. It'll explain how love is both intensely private but also made public by the rules our society imposes about how and who we love. It'll explore the loves we underestimate and ask you to reconsider love not as an emotion but as a need as fundamental to us as the food we eat and the air we breathe. And it will touch on the less considered aspects of love;

its darker sides and where our quest for love might take us in the future. It is my fervent hope this book will act to both reassure and challenge you. As humans, the outlets for our love are so many that I truly believe that we can all find love in our lives, be it with a lover, friend, dog or god. But the question remains, are we able to sit comfortably with a phenomenon that can both heal and harm, and which, ultimately, is guaranteed to be unpredictable?

I am writing this prologue during the second Covid-19 wave in the UK. Covid has been devastating to us all on so many fronts, but I think one good thing that may have come out of it is a renewed understanding, through experience, of what is most important in our lives; to our health, our happiness, our life satisfaction. And it is who we love. Because Covid has taken away our opportunity to be with each other and has brought to the forefront our immense, visceral *need* for each other, whether it be the hug of our friends or parents, or the key workers who have made sure we still receive the essential elements of life; food, water, care. People who have perhaps sacrificed their contact with their loved ones – as many in our health service have – to ensure they are present to look after the loved ones of others. Human cooperation, human love is awe-inspiring. I believe it defines our humanity. And Covid has shown us that when everything else is stripped away it is all we have and, ultimately, all we need.

But first. To start at the beginning. Love as survival.

A note about my interviewees

Love is hugely subjective. How I love and how you love are very likely to be different. As a consequence, no account of love, I believe, is complete without hearing the opinions and experiences of other people. Yes, we have more of an objective understanding of love than ever before, but there are some questions to which the answers cannot be found on the scanner screen or at the bottom of the petri dish. So my work always involves talking to real people and collecting their thoughts. This book is no different, and throughout it you will hear the anonymised experiences of those who may have experienced a certain form of love – including polyamorists, aromantics and nuns – alongside the more general thoughts of many members of the public at large who responded to my Twitter pleas with an openness that has been overwhelming. In a time of lockdown and social distance, it has been a joy to connect with so many through the screen, and their stories have often brightened my day. I have asked all to tell me their definition of love and to share their experiences of the love in their lives. For some this has been a joy, for others a more difficult experience and to them I say a particular thank you.

CHAPTER ONE

SURVIVAL

● ● ● ●

'Love and compassion are necessities, not luxuries. Without them humanity cannot survive.'

Dalai Lama

'There is nothing on this planet that's more annoying than another person.'

Dr Stan Tatkin, relationship therapist

To love is to survive. I'm not talking about the desperation of the teenager who is convinced they will *die* if the object of their affection doesn't at least start to acknowledge their existence. Or the belief of the spurned lover that the pain of their broken heart – surely a sign of imminent death – can only be mended by a swift reconciliation. I'm talking about actual fundamental survival, the stuff of 'are you going to be sending some genes forward down the generations or not?'-type survival. Because humans are tricky beasts. We have massive brains which have enabled us to create, to explore, to conquer and innovate, but they have also meant that we can't actually reproduce effectively without considerable input from other people, nor can we learn everything we need to know without resorting to the help of friends, parents or Google, itself a fount of human-generated wisdom. Whether we like it or not, we need each other. At the basis of all relationships – close or merely passing – is cooperation. And decades of study have shown us that humans are arguably the most cooperative species on the planet, so extensive is our social network, so diverse the members of it and so complex and inter-related the relationships within it. Love stems from cooperation and cooperation *is* our route to survival. To understand love today, we must understand why it evolved, and that means we must begin with an understanding of cooperation, both why it is critical to our survival and why it can be, at times, a massive headache.

The Reluctant Parent

I wasn't really one for human babies and children as an adolescent. While some of my friends became incoherent with joy on encountering a baby, any kid had to have much more fur and, ideally, paws before I wanted to get involved. I had a plethora of pets as a child but my first real experience of caring for a youngster came when I was completing my masters research in primatology at London Zoo. I was studying foraging behaviour in a troop of Sulawesi Macaques, attempting to work out the evolutionary origins of the much-discussed sex difference in human navigation ability, the basis of many an un-PC comedian's jokes. As well as the species bearing a welcome resemblance to a be-quiffed Elvis, this troop served up a daily diet of drama – arguments, reconciliations, bids for power and, most importantly for this wannabe monkey mum, births – all of which tended to be much more exciting for me to watch than the carefully controlled task I had set my simian participants. Generally, the mums in the troop had mothering pinned down. They would often give birth at night, unaided, and would then go forward to care for their new-born without much help from anyone else. There was an initial fascination with this new group member – everyone would have to have a quick hold or groom – and every now and then a younger female might pinch a baby for a bit of practice, much to the outrage of mum, but generally raising a macaque baby was a one-monkey job. And macaque babies, as with all the non-human primate species, are quickly independent, moving around and exploring their enclosure and playing with fellow youngsters only a matter of weeks after birth. Mum would only be returned to for food or a carry when play became too exhausting. However, one of our mums, Mia, struggled with mothering. Having lost one baby due to neglect, when she gave birth a second time the keepers kept a close eye on her and, unfortunately, history repeated itself. Mia seemed uninterested in caring for her baby. Because adoption is a rarity in the animal kingdom, the keepers had no choice but to step in and hand-rear the baby themselves. Hence my first experience of mothering came courtesy of a furry baby with the biggest eyes and the tiniest fingers and an insatiable appetite for clambering around the staff room, exploring all its nooks and crannies, while we had our morning tea.

Flash forward ten years and you find me with my firstborn. Now, I have read all the books and attended the classes, I am a one-woman knowledge machine, but my experience of being a mum is very different from that of a macaque mum. My baby is unbelievably helpless when born. She cannot focus her eyes or coordinate her limbs and requires active intervention to feed, burp, fall asleep, entertain herself and, most onerous of all for the weak of stomach, to clean herself once she has defaecated. She won't be able to lift her head until four weeks, master hand-to-mouth coordination until sixteen weeks, babble until twenty-four weeks and sit independently until thirty-two weeks. She will only start to play around six months and it may take her until she is two years old to walk. Beyond this, she will have to have input from a whole team of adults to help her navigate her childhood and adolescence, in the form of family members, friends, teachers and medics. She will benefit from the knowledge of teachers, from the protection of medics, from the support and challenge of peers and from the care of her family. Without all this, she would be very unlikely to survive, let alone thrive.

An Anatomical Anomaly

My mum friends are my key friends because we are all at the same stage, mid-thirties, we don't have loads of time. We have shared interests. 'Oh, mine isn't eating', you know? There's a comfort thing there. **Joan**

Compared to our macaque baby, human babies need *so* much input from a myriad number of players because of an evolutionary quirk which has meant that our babies are born far earlier than they should be. This quirk is caused by the unique combination of a massive brain – it's six times bigger than it should be for a mammal of our size – and our mode of walking – on two legs. The result of this coincidence is that if a baby went full term her head would not fit through our narrowed birth canal, mum and baby would die and the species line would come to an abrupt end. So we have evolved to birth our babies very early, resulting in a baby whose brain is not yet fully developed – and hence is incapable of doing anything alone for a significant period of time post-birth – and a mum who needs help to care for, and wrangle, her ever-growing horde of helpless babies and wayward toddlers. In the macaque world, because babies are born developmentally well advanced, a mum only has to actively care for one offspring at a time. She will make sure one baby is fully off her hands before the next one puts in an appearance. But this is not the case for human mums, as many a harassed parent will attest. Human children are dependent upon their carers for many years. My children are entering their teenage years but their need for constant input has not abated. My mum and dad frequently argue that at the age of forty-five I am still a cause of worry and stress. Add to this the fact that our capacity for technological innovation appears to develop at warp speed and the result is a child who needs the input not only of carers, but teachers too, to make sure that she survives and thrives as an adult in an increasingly complex world.

Girl Power

In the first instance, mum would have turned to her female kin to help her with raising her unruly horde. Turns would be taken babysitting as survival-critical food and water were sourced, older mothers would teach younger ones the key skills of child care, and group members would ensure teenagers (a life stage unique to our species) were fully versed in the latest technological innovations in hunting and fire production, and introduced to the subtleties of social politics. Cooperating is nothing if not a political labyrinth. As with the majority of mammalian species, dad was nowhere to be seen. But about half a million years ago our brains expanded again and suddenly baby was even more dependent when born, and took even longer to develop. The help of just your mum, aunts and sisters was no longer going to cut it. As a consequence, evolution caused the investing human father to evolve to pick up the slack and to make sure we didn't become an evolutionary dead end. (For the full story of his evolution, see my book *The Life of Dad*.)

The Battle of the Sexes

The arrival of dad presented a whole new set of issues around cooperation. While cooperating to raise children with your female kin was largely an issue of trading like with like – namely childcare favours – cooperating with the other sex was a whole new ball game. Rather than being altruistic caregivers, dad wanted to help raise the offspring so he could make sure he was readily available as mum's next sexual partner, and that any children going forward were definitely his to invest in. This sex-for-childcare exchange is cognitively *so* much more complex then swapping childcare favours with other females, because you are dealing in different currencies. This results in a complex exchange calculation to make sure no one comes out on the wrong end of the deal. Hence while dad was a welcome addition to the childcare team, his arrival did mean that we had to invest even more precious time and brain power in maintaining this cross-sex relationship. The human cooperation network was getting increasingly complex.

Alongside childcare and teaching, we are still missing one of the most vital benefits of our cooperative network, without which we would not survive for days let alone be around to continue the species. We have to cooperate to subsist. In the environment in which we evolved, this may have meant relying on others' knowledge to learn the skills of hunting, or to locate water sources, working together to build shelters and forage for food, and trading favours with specialists so they would produce you a new arrow head or hunting spear. Even in our modern world where you can order your groceries from the comfort of your sofa to be delivered straight to your door with barely any actual face-to-face interaction, just think of the number of people who are involved in growing, harvesting, conveying, packaging, picking and delivering your order. You are cooperating with them all, albeit at a distance. We all must cooperate to survive. To learn, to raise children, to eat. And to achieve this end we must cultivate an extensive and complex network to make sure all of our survival-critical bases are covered.

The Power of 150

One of things I realised with this blasted lockdown was that I don't have as many friends as I thought I did! I was surprised by how few people got in touch and how many people . . . well, almost no one reached out to see if I was OK. That made me think about friends a lot. I realised I have a lot of acquaintances who I might be fond of but I don't really love. I don't have a fundamental relationship with them. I realised I only have one friend. **James**

The result of this need to cooperate with each other to breed, learn and subsist, often repeatedly, is that we build a network of relationships. What is fascinating about this network is that regardless of age, personality, gender, ethnic background or any number of possible individual differences, we all interact with the members of this network, which is organised into distinct layers, in a broadly similar way. So those who sit alongside us at the centre of this network are the four or five people to whom we are emotionally closest and who we interact with most often, at least weekly. This is the 'one friend' and probably a few family members – generally a spouse and children – who James refers to above. These are the members of an exclusive club which we at Oxford denoted the Central Support Clique. They may be our parents, partner, children or best friend. They are the people who you turn to at your most emotionally difficult times in the sure knowledge that they will respond. In the next layer of the network are the fifteen or so people who constitute the Sympathy Group, so-called because research by psychologists Christian Buys and Kenneth Larsen in the late seventies found that this is around the number of people we can maintain intense relationships with and feel genuine sympathy for. More practically, these are those who you go out with for the night, to the pub, cinema or restaurant; your party crowd. After the Sympathy Group we move to the forty-five layer, the Affinity Group. This is generally the home of extended family, acquaintances and some work colleagues. With the next layer we reach the limit of the active network at 150 – the people with whom you share a history but who you may see only once a year. The layers continue, to 500, 1500, 5000 and beyond. What you may have noticed is that these layers, as you move away from you – ego – at the centre, increase in size on a scalar of roughly three, with each layer being inclusive of the ones within it. From above, it looks like a set of concentric circles, centred on you; a dartboard where you are the bullseye. Here's a figure to help you visualise it.

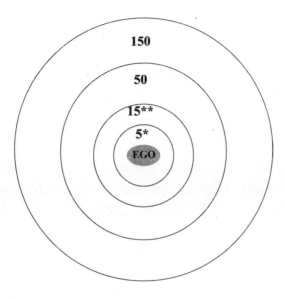

The social network after Robin Dunbar
Image courtesy of Robin Dunbar

Beyond the 150 layer, all else is at most mere acquaintance. Members of the 500 layer we may be able to name and know personally, while those in the 1500 layer are nameable but include those we have never met, such as celebrities or politicians. So, whilst you might not know them personally, the Queen or the President of the United States are still in your network. Within the 5000 layer we recognise faces but no names. This network structure, and this active outer limit at 150, is so consistent between people because it is constrained by two key factors: the time we have to devote to our relationships and the cognitive resources we bring to them. We all push our capacity for social time to the limit so the 150 is the result of the *maximum* time and brain power we can commit to our social world. Our time is finite, our social budget must fight for space with all the other essentials of life: work, food, rest. But not all relationships are equal. We devote 40 per cent of our social time to the five people who sit alongside us in the centre of our network, with the next layer of ten people getting the next 20 per cent. Being social uses considerable brain power both to keep track of who has done what, to remember who everyone is and crucially the history you have with them, to stick to the rules of social interaction, including turn taking in conversation and inhibiting your less helpful responses and behaviours, and to spot a cheat (more on this below). Indeed, a large portion of our conscious brain, the prefrontal cortex, is given over to the job of being social, meaning that other areas, such as areas dedicated to olfaction (smelling) have been drastically reduced as compared to our fellow mammals. And because there is only so much brain power we can commit to this and be able to function in the rest of our life, the limit of the active network is stable at an average of 150.

This number of 150 has been found so consistently in the data on human social behaviour that it has a name: Dunbar's Number, named for my boss at Oxford and its discoverer, Professor Robin Dunbar. Robin, who is something of a guru among the social-media entrepreneurs of Silicon Valley, has collected data on social interactions from groups of people as diverse as European mobile-phone users, African hunter-gatherer tribes, factory employees and Viking sagas. While the range of network size tends to vary from 100 to 250, the average is always 150. So take the range of units in the modern army. Layer 1 of the social network is a special-forces unit (~5 individuals), layer 2 is a section (~14), layer 3 a platoon (~45), layer 4 a company (~150) and Layer 5 a battalion (~300–800). These sizings may have been arrived at by trial and error concerning what works in the field of action over many years, but they exist *because* this model of social organisa-tion is the one that gives the best chance of survival based upon the strength of bond and the speed of communication required in a particular context. This social-network arrangement is an adap-tation arrived at by natural selection just like any other evolved trait. The consequence of this is a stability of numbers which has endured from the Vikings (and probably before) to us.

Variance in Dunbar's number between individuals is the result of differences such as age – network size tends to peak in our twenties and diminish in old age; personality – unsurprisingly extroverts have bigger networks than introverts; and sex – women tend to have bigger networks than men. In a study by Robin and our collaborators based at Aalto University in Finland, analysis of the mobile-phone records of 3.2 million people showed that individuals spent nearly seven times as long on calls to people in the inner layer of their network as the average for members of the wider network. For both men and women, the size of the social network peaked at the age of twenty-five, although at this point males had more connections than females. From this point onwards, size diminishes until at the age of around thirty-nine women start to have larger networks than men, and there is in fact a second peak in size for women at the age of fifty. This is in-teresting as this is the average age of menopause and this increase in social connections may mark the freeing up of time caused by mature children flying the nest and allowing women to spend less time caring and more time investing in other relationships. A particularly interesting finding about women's networks is that they tend to have more people in their inner circles, so more close friends, as compared to men, and they invest more time in them. We'll look again at the importance of close friends to women – and the love they have for them – in Chapter 4.

A Little Bit of Shakespeare

Cognitive ability is also a major factor in individual difference. People with larger prefrontal cortices – the area of the brain that is largely responsible for our social cognition – and who have denser white matter – enabling swift communication between the areas of the brain – tend to have larger social networks as they are better able to keep track of what everyone is doing. At the most basic level, theory of mind, the ability to second-guess what someone is about to do is critical for this. However, due to the complexity of our social networks – we are connected to everyone in our network but they are also likely to be connected to each other – our need to understand the mental states of others does not stop there. We can also understand situations which can be described thus:

I understand that John believes that Mary knows that Stuart imagines that Jane is cheating on him.

The text in italics is theory of mind, or second-order intentionality – required in a direct relationship between two people. The whole sentence constitutes fifth-order intentionality and enables us to understand the states of mind of people who are not in a direct two-way relationship with us. Indeed, we may just be inferring their intention from a story which is being told about them.

We all vary in our ability to understand the intentions of others, known as mentalising. The average person can handle the fifth-order intentionality represented by the sentence above, with the range of ability running from a minimum of third (that's 'I', John and Mary) to as high as seventh order (we would need to add two more people to our story) in a rare few people. Apparently his plays show that Shakespeare was a bit of a mentalising whizz – he could do sixth order – which is why his plays have so many complex, interrelated relationships within them and why those of us with lesser mentalising skill may struggle to follow them. The ability to mentalise is obviously vital to spot cheats, anticipate when it is your turn in a group conversation or predict how your behaviour might impact others in your network, but it is also crucial for the key lubricant of social interaction, language. We rarely say exactly what we mean, instead relying on shared knowledge of in-jokes, metaphors or turns of phrase, meaning that we have to use intentionality to interpret what the other person means to say. And research has shown that there is a direct, positive relationship between a person's mentalising ability, their brain size and the size of their social network. In a study run by Robin Dunbar and colleagues at Oxford, they found that not only did neuronal activity increase as the number of levels of intentionality in a series of story vignettes increased, but there was a direct relationship between the size of the orbitofrontal cortex, the part of the prefrontal cortex just behind the eyes, and the maximum number of levels of intentionality a person could handle. In turn, the size of this brain area correlated positively with the size of the person's social network. How amazing is that?

So, we must cooperate; to subsist, to learn and to raise our children. And because of the universal constraints placed on our network by time and the cognitive abilities we bring to this vital area of our lives, we all tend to interact with each other in a broadly similar way, hence we have a recognisable and replicable network structure, and a maximum active size at around 150 people, Dunbar's Number.

Oh to Be an Island . . .

The problem with all this need for each other is that people lie, and they cheat, and they steal. You have to be very good at spotting these people if you are going to avoid expending precious time and energy on relationships which at the very least may harm you emotionally or financially, but may at the extreme have the potential to threaten your health or very survival. At a minimum, you need to employ theory of mind, if not higher levels of intentionality if the betrayal involves more than one person, to make sure you don't get taken for a ride. In an ideal, and blissfully solitary, world, you would invest this neural energy in something else. Add to this the stresses of group living – the competition for resources, the need to coordinate your day with others rather than having the absolute freedom to do whatever you want, or the necessity of living in a hierarchical society. Like most group-living primates, we exist within a strict hierarchy which is based upon a combination of our attractiveness, our wealth and our status, which are, as with everything, proxies for the likelihood of reproductive success – the ultimate measure of a successful life from an evolutionary viewpoint. Existing in a hierarchy is a stressful and time-consuming activity. If you are at the top, you have to spend time and resources maintaining your position – currying favour with allies, displaying your vast wealth/attractiveness and fending off all those who would usurp you. If you are at the bottom, then you are the last to gain access to all resources, which can mean thirst, starvation and no opportunity to reproduce. And if you are in the middle, you are arguably in the worst place (at least if you are at the bottom, those at the top tend to not think you are worth bothering with).

You have those above you stamping down as they work to prevent your rise and those below pushing up as they try to leapfrog you to the top. You really are the squeezed middle. All of this manoeuvring means that we are all constantly monitoring each other, which taxes your brain – your mentalising abilities will definitely be tested – and takes your time. And last but not least, let's not forget the unique stresses and costs of cooperating with the opposite sex as we struggle to raise our children. The trade-off between men and women hasn't got any easier to deal with in the last half a million years.

The consequence of all this is that cooperation is at once both critically necessary and life-threateningly stressful. So, what has evolution come up with to ensure that we cooperate to survive and reproduce despite the difficulties involved?

Love

At its most basic level, love is biological bribery. It is a set of neurochemicals which motivate you to, and reward you for, commencing relationships with those in your life who you need to cooperate with – friends, family, lovers, the wider community – and then work to maintain them. As we will see in the next chapter, the sensations which these chemicals induce in the individual – and which we call the sensation of loving or at least liking – are there to make you feel warm, content, euphoric and encourage you not only to seek out new sources of this sensation but also motivate you to keep investing in your relationships in the long term so that the feeling, and the survival-essential cooperation, never ends.

Love: The Route to Health and Happiness

Who am I really, in isolation? I am always in relation to other people. So there is something about the people when you are with them. They are bringing out your best self. Your happiest self. The person I most enjoy being. When I am with them there is a sort of lifting of 'Oh, not only am I feeling this joy of being with you but I am feeling the joy of being allowed to be this version of me.' There is a self-love that happens when you are with someone else you love that you can only get by being with them. **Margaret**

I am sure we can all imagine how critical we were to each other in the knife-edge environments of our evolutionary past, and there are certainly areas of the world today where having the cooperation of others is still the difference between life and death. But surely here in the west, where our environment is relatively benign, and the service sector has seen fit to try and make everything we need to survive accessible from our sofa, cooperation, and in particular our closest relationships, are less about survival and more just about fun and belonging. We know what the important things for a healthy life are: exercise, a balanced diet, not smoking and maintaining a healthy weight. That's it. We have survival cracked. But a seminal study carried out by psychologist Julianne Holt-Lunstad and her colleagues in 2010 would beg to differ. Julianne collected the data together from 148 studies which had explored rates of mortality following chronic illness – cancer, cardiovascular disease and renal failure being the most prominent – and aspects of an individual's social network. For some studies this was the size of their network, their actual or perceived access to social support, their social isolation or loneliness, or the extent to which they were integrated into their network. Having carried out some very complex statistics to ensure she was comparing like with like, she concluded that being within a supportive social network reduced the risk of mortality by 50 per cent. That places it on a par with quitting smoking, and of more influence than maintaining a healthy BMI measure.

My friends bring a support system that I know I can rely on. There is a dependability with them that I can rely on regardless. If I need cheering up, I know I can go to Bruno. If I need advice, career advice, I'll go to David. If I need emotional, mental-health conversations, I'll go to Nick. They provide similar but different attributes that I know I can go to. Surrounding myself with this support system means that whatever trouble or difficulty arises, I have support. **Doug**

Since Julianne's study, numerous other projects have reinforced this conclusion; that having good-quality social relationships (known as social capital) is *the* most important factor in your health, happiness and life satisfaction. Indeed, in 2019 a group from Harvard in the US, led by Justin Rodgers, repeated Julianne's study with the body of social-capital and health research published in the period 2007 to 2018. After reviewing 145 studies (in fact 1608 articles were published in this time but not all made it through the robust selection criteria), the Harvard team concluded that your social capital – be this the size or cohesion of your social network, your level of reciprocity or participation, your levels of trust, belonging or rate of volunteering – had a significant impact on your overall mortality or life expectancy, your risk of dying from cardiovascular disease, cancer or diabetes, the likelihood you are obese and your perception of your own health. As I write towards the end of 2020, studies finding a link between social capital and cognitive function in the elderly, adherence to HIV pre-exposure prophylaxis in at-risk gay men, reducing the risk of poor mental health following the acquisition of a disability and self-perception of health have been published. The question arises as to why being in good relationships has such a marked impact on our health? The reason is multifaceted, but explanations include the simple fact that having friends and family brings helpful resources such as money, practical care or health knowledge; that they make you feel better psychologically, which reduces the impact of stress on your body and improves your mental and physical health; or, most tantalisingly, that the neurochemicals which are released when you interact with those you love have a direct role in promoting the efficient functioning of your immune system.

'Well, Hello, Beta-endorphin!'

I always feel better when I have seen my friends. So I saw one of them yesterday . . . I don't get funny but I feel 'Hmmm, I haven't seen anyone for a couple of days.' You get to offload . . . I need the balance of all my different friends. So mummy friends but also friends who I talk about books with and where we want to go. It is cathartic and we laugh. Life is busy and if you keep it all in your head it is unhealthy. **Joan**

We will learn in the next chapter that the sensation of love is underpinned by a cocktail of neurochemicals which are released when we interact with our friends and family. One of these neuro-chemicals – and the one I argue is the key to our ability to love in the long term – is known as beta-endorphin. Some of you may know this as the basis of your body's natural pain-killing system or the source of the euphoric feeling which follows a bout of vigorous exercise – the phenomenon of the runner's high – but it also appears to have a key role to play in the operation of our immune system. In 2012 endocrinologist Dipak Sarakar, who is based at Rutgers University in New Jersey, published his findings, based upon research in rats, that the mu-opioid and delta-opioid receptors had a role in the function of the natural killer cells which make up part of the mammalian immune system, ours included. The mu-opioid receptor, in particular, is the receptor in the brain upon which beta-endorphin acts, and as such Dipak's work allows us to suggest that the release of beta-endorphin during social interaction stimulates the natural killer cells, meaning that unwanted pathogens are dealt with more efficiently than if social interaction has not occurred. This study still needs to be replicated in humans – the knocking out of some relevant genes in the rats makes this a tricky goal to achieve – but Dipak's work offers the tantalising possibility that social interaction has an integral role to play in the operation of the body's defence systems.

I hope it is clear by now that, whether we like it or not, we need each other and that love is the force which motivates us to overcome the difficulties of group living to cooperate at a level unmatched by any other species. We must cooperate to subsist, to learn, to raise our children, to innovate and create. We build complex and enduring networks encompassing our families, our friends, our co-workers and our lovers, which, regardless of individual differences, all follow the same pattern. Beyond the water, food and shelter that we need just to survive, our relationship with those we love has the largest impact on our health and happiness, our life satisfaction and longevity. Love has been around a long time but it is still as much about survival today as it has always been.

CHAPTER TWO

ADDICTION

● ● ● ●

*'At every stage, addiction is driven by one of the most
powerful, mysterious, and vital forces of human existence.
What drives addiction is longing – a longing not just of brain,
belly, or loins but finally of the heart.'*

Cornelius Plantinga, theologian

Lucy is a drug addict. Her thoughts and daily plans are consumed
by the need to score her next hit. She ignores the opinions and
concerns of others as her obsession overwhelms her – she is
blind to the reality of her situation. The possibility that she may
not be able to satiate her desire leads to a mental and physical pain
that paralyses her as she withdraws. All else pales into insignificance next to the relationship she has with her drug: appointments
are missed, meals abandoned, friendships lost, work becomes an
unhelpful distraction. But a hit of her drug leaves her in a haze of
euphoric bliss and contentment; her world is complete. Heroin
addiction controls her life. Just as love controls yours.

In 1983 the American psychiatrist Michael Liebowitz published
a book entitled *The Chemistry of Love*. In it he drew parallels

between his experience of treating opioid addicts and the behaviours exhibited by those who are deeply in love. He pointed to the euphoric highs, the intense cravings for satiation which consume all our attention and the physical and emotional pain of withdrawal. His work was based on nothing more than observation and anecdote, but his belief that the 'drugs' of love were akin to those which fulfilled the cravings of a drug addict was the prompt that neurobiologists needed to begin their search for the neurochemistry of love. And that research shows that we are indeed addicted to love.

In this chapter I will reveal that the biological bribery which I introduced in Chapter 1 takes the form of a set of neurochemicals which motivate us to commence – and then maintain – the relationships that are key to our survival. These neurochemicals make up a heady cocktail, the ingredients of which alter as we move from attraction, and possibly lust, to love. Of course, oxytocin is in there, but it is joined by dopamine, serotonin and beta-endorphin, all of which have an equally important role to play. And while the individual characteristics of different forms of love influence the fine detail of our brain activity, there *is* a basic neural fingerprint of love that all share to ensure we experience love at both the conscious and unconscious level.

The Birth of a Relationship

'A conversational chemistry draws me to people. And the little hints there could be more . . . if things go a bit deeper than the shared facts of existence and you start to pick up very quickly on values. And that brings you to ask do I want to see this person again or has it just been an entertaining chat at a dinner?' **Marie**

While lust is confined to the sexual relationship, attraction is a stage which defines the beginning of all our close relationships, including those with our children and our friends.

In the very first moments of meeting – be this in the wine bar, the birthing room or the school yard – oxytocin and dopamine act in partnership to give us the confidence and motivation to approach the other person and begin the first stages of bond formation. Oxytocin is a neurochemical, produced in the brain, which also has a physiological role in the body, particularly with respect to childbirth and breastfeeding. However, over evolutionary time it has been co-opted by those areas of our brain linked to our social behaviour to play a key role in the establishment of our relationships. Within the brain, oxytocin works to lower our inhibitions to forming new relationships. It does this by 'quietening' or de-activating the amygdala, the tiny almond-shaped structure at the very core of our unconscious brain, where fear sits. So that nagging voice at the back of your head which tells you that you are sure to be rejected by your potential new lover and have to take the walk of shame back across the bar in front of an intrigued audience is reduced, giving you the confidence to take the first step. You are left with an ethereal sense of calm and capability.

'I do love my friends. Why? They both lift me up. You get that dopamine hit, that kind of rush of adrenaline and that excitement when you see them. I always come away from in-person hanging out with them lighter. They fill me up.' **George**

However, if oxytocin were to act alone, it might have the impact of making you feel *so* chilled that you failed to make it off the bar stool or make the effort to meet the needs of your crying baby, which is not good news if we are to continue the species. Here, the relationship between the release of oxytocin and dopamine is key. Whenever oxytocin is released, dopamine accompanies it, and as many of you probably already know, dopamine is our brain's general reward chemical. It is released when you do anything that you perceive to be enjoyable. For me that is consuming banoffee pie, drinking gin and tonic and hugging a dog or two, preferably all at the same time. This is the chemical which is released when your Instagram lights up with 'likes' and your self-esteem swells. But dopamine is also released when you see someone you are attracted to, and its role is not only as a reward but to give you the kick you need to make sure you actually make the effort to connect with the person. This is because dopamine is also known as the hormone of vigour. It is wired into your motor circuits and as well as giving you an enjoyable chemical reward for making the effort to be social, it is also the reason why we persist in pursuing new relationships despite the effort that may be involved. An accumulating body of evidence has shown us that the partnership between dopamine and oxytocin means that you have access to the motivation to make a move and the drive to focus that attention on forming a new relationship.

[My husband and I] fell in love so immediately; in the first few weeks I remember feeling so elated and excited, but at the same time I had this deep sense that this person wouldn't let me down. It was so equally reciprocated between us . . . I remember in the early days I would wake up and think 'Wow, this is really happening to me' and I couldn't quite believe it. Our love has mellowed and developed, but it still feels like a soft cocoon, a place where everything is OK with the world. **Jayne**

And there is strong evidence that those who are newly in love do have higher levels of circulating oxytocin than those who are single. In their 2012 exploration of the role for oxytocin in the early stages of love, neuroscientists Inn Schneiderman, Orna Zagoory-Sharon, James Leckman and Ruth Feldman carried out a longitudinal study comparing the baseline oxytocin levels of a group of new lovers and a group of singles. As they predicted, they found that the new lovers group, sixty heterosexual couples who had been together for three months, had significantly higher levels of circulating oxytocin than the singles. But more fascinating was that their results suggested that by knowing a couple's baseline level of oxytocin at the start of the study they could predict whether or not they would still be together six months later. This slightly scary finding, which definitely removes some of the mystery surrounding love, showed that of the thirty-six couples who were re-tested at six months, the twenty-five couples who had stuck together were those who had showed the highest oxytocin levels at the start of the study. Those couples who had been at low levels had not survived the course.

As well as being the glue which makes the early stages possible, oxytocin and dopamine are also critical at this time because they enable you to start to memorise the key details about your new love. What they look like, how they sound, how they smell, what they like or dislike. Think of the focus of a new father or mother on their newborn baby's perfect fingers and toes as they encode the image into their brain, or their ability to pick out their baby's cry from the cacophony of the maternity ward. Dopamine and oxytocin achieve this by increasing the plasticity of the brain and making it more open to change. The brain is more able to reorganise itself around the new relationship, allowing the new person to be quickly and efficiently incorporated into our sense of self. The result is that memories are built, and our attention becomes orientated in the direction of our new love.

Obsessive Love

The need to pay our new love attention in a busy and distracting world is where the third neurochemical in our cocktail comes into the mix. Serotonin is the chemical which regulates your mood, happiness and anxiety. Low levels are implicated in a range of mental-health conditions, including depression and obsessive compulsive disorder (OCD). And it is its role in this latter condition that gives some clue as to its possible role in love. When you are first attracted to someone, while your levels of oxytocin and dopamine rise, your levels of serotonin drop. While the jury is still out on exactly what serotonin's role is, it seems increasingly likely that this drop in serotonin underpins the obsessive element of love, as low levels are found in those who suffer from OCD. This change underpins your desire to share photos of and anecdotes about your newborn with anyone who will listen, the tendency to eschew work in favour of daydreaming about your new lover or the enthusiasm with which you embrace a new best friendship. We have to be vaguely obsessed with those with whom we have relationships to ensure we make the effort to coordinate our lives with theirs, to focus on their needs and to remember to make time for them. In a way this reaches its peak in the human family. I definitely need a measure of obsession to confront the epic challenge of coordinating the activities of my kids, husband and me, to ensure everyone's individual dietary quirks are catered for and be ready to listen to *everyone's* stories of woe or joy at the end of the day. Sometimes solitude on an island is very tempting.

Obviously in extreme cases this obsession can have a darker side, as it becomes pathological obsession, something we will explore in Chapter 9, but for most of us a healthy reduction in serotonin is vital to keep the wheels of our relationship on the road.

The partnership of oxytocin and dopamine is undoubtedly critical during the establishment of a new relationship, but they are aided in this critical phase by our senses. The early stages of attraction are largely unconscious and, as with the lesser mammals, the senses have a key role to play. This is particularly striking when we consider the first stages of romantic love; more accurately identified as 'lust'. Whether or not we get the first stirrings of lust when we lock eyes with someone across a crowded room or train carriage is largely down to the results of a complex algorithmic calculation undertaken by our brains in response to the input of our senses. The key senses at this stage are the senses of sight, hearing and smell. Touch and taste come later, as your intimacy grows, and to understand whether or not a person gets an encouraging 'yes' or a dismissive 'no' you have to understand a little bit about the world of human dating and mating.

Welcome to the Dating Game

The human mating game is based upon a competitive market akin to the stock market, but rather than our worth being expressed in pounds, euros or dollars, it is expressed in mate value. We each have a biological mate value on our heads and this value is calculated based upon the likelihood that we will be reproductively successful; that is, that we will be able to produce viable offspring and raise them successfully to adulthood, when they themselves will reproduce. The more times we can do this, evolution has decreed, the higher our value. Some of us are more likely to be able to do this than others. How this value is calculated differs for men and women because of the different roles they play in the reproductive game (here I am talking about heterosexual attraction, I consider homosexual attraction later on in the chapter). For women it is a calculation based upon their health, fertility and, if we are looking at a long-term relationship, their fidelity. Evolution wants women to be able to get pregnant, carry that pregnancy to term and be healthy enough to live to raise the child. For men it is all about their ability to protect, provide and commit to their family – remember, in the world in which this system evolved, women were either pregnant or breastfeeding constantly, so were incredibly vulnerable. For both, genes are also a factor, although less so if the relationship is to be long term. To be able to make this calculation, your brain uses your senses to take in key indicators of these attributes. And while we all differ to a certain extent in what we find attractive – heaven forbid anything we do is straightforward – there are some general trends that we can identify from our numerous scientific observations of people's dating behaviour.

It's All in the Eye of the Beholder

Take a woman's waist-to-hip ratio, one of the most robust indicators of health and fertility. Cross-cultural studies of female body shape have repeatedly shown that the most attractive waist-to-hip ratio is 0.7: the classic hourglass. This might surprise you because of our overwhelming focus in the west, in particular, on thinness, which would suggest that large hips are not ideal. But there are two things to say about this. First, it is the *ratio* of waist to hips that is key, so you can be a size 8 or 18 and the 0.7 is still what is important. Secondly, the obsession in the west with thinness is largely media driven and doesn't reflect what people actually find to be attractive. In fact, the 0.7 is culturally stable – its link to attraction has been tested in countries from Indonesia to America – because of its strong connection to general fertility and several positive health outcomes. In his 2002 paper summarising the evidence for the power of a 0.7 ratio, American psychologist Devendra Singh concluded that those with a 0.7 ratio had a lower risk of cardio-vascular disease, adult-onset diabetes, hypertension, endometrial, ovarian and breast cancer, gall-bladder disease and early mortality. This is a consequence of the link in women between the sex hormones and the distribution of fat. Further, because both pre-puberty and post-menopause fat is distributed along male lines, leading to a high waist-to-hip ratio, a 0.7 ratio is a good indicator of fertility and fecundity. Why does all this matter? Because if a man is to remove himself from the dating game, he wants to make sure that the woman he chooses is both sufficiently fertile and adequately healthy to conceive, carry and raise a child to maturity, ensuring the perpetuation of his genes down the generations. Of course this focus on 0.7 does not mean that men *only* find this ratio to be attractive – obviously men choose women with a wide range of ratios – but it is the *most* attractive and, therefore, all else being equal, the goal.

And the importance of the WHR to male mating decisions is confirmed not only in the lab but also in the real world. In their 2015 eye-tracking experiment, Ray Garza, Roberto Heredia and Anna Cieslicka from Texas A and M International University showed that men, and interestingly women, focused first and spent the longest time looking at an unknown woman's midriff before moving on to her face. This would suggest that when deciding who to approach as a potential mate, or who is the competition if you are female, the waist-to-hip ratio is one of the first pieces of information processed by our brain's algorithm.

Beyond the waist-to-hip ratio, men are assessed on their shoulder-to-waist ratio and their height. When it comes to the shoulder-to-waist ratio, a 1.4 is the goal, although this is rarely displayed by anyone other than an Olympic athlete or serious gym junky. But this figure is tied to athleticism, which suggests the ability to dominate and protect. When it comes to height, tall, but not too tall, is the goal. Tall men are perceived to be more dominant and more successful, and there is evidence to show that this perception is often correct. Studies have repeatedly shown that tall men are more successful in their careers, meaning that they bring more resources to the relationship.

For both sexes, the face, and more accurately its degree of asymmetry, is a vital source of information when assessing the value of a potential mate. This is because there is a close tie between symmetry and healthy genes. Why is this? As with many other animals, the human body is bilaterally symmetrical: two legs, arms, feet, hands, eyes and ears. Given unfettered influence on our developing morphology, genes are designed to make us completely symmetrical. However, pure symmetry does not exist in nature because during development the genes encounter many environmental challenges which push them off their course. However, if your degree of facial asymmetry is low – meaning you are closer to pure symmetry – it shows that your genes are strong enough to deal with the environmental challenges and still do a pretty good job on the symmetry front. For humans, and particularly women, the degree of asymmetry is of particular importance if a short-term relationship is on the cards as, in the absence of long-term commitment, resources and protection, the genes are the only thing she will be getting from her mate. Hence the trend is for women to focus more on 'good looks' in a short-term relationship than when selecting their long-term mate.

The Sweet Sound of Love

As with sight, hearing also enables us to collect a lot of information about our potential mate. Voice pitch is critical for both sexes, with women preferring a lowered, though not too low, male voice as it indicates large body size and a good level of testosterone, both of which imply a man who is of high status and can protect. Indeed, men are seen to unconsciously lower their voice pitch when in the presence of a woman they fancy. In contrast, men prefer a high voice as it is perceived to be feminine. But beyond the pitch of our voice it is what we say that is most attractive. The human brain is arguably the sexiest organ in our body because it enables us to use language creatively, to be humorous and to produce great works of art or music. These are all indicators of our cognitive flexibility, which is an attribute that we would all like our offspring to inherit as it is linked to intelligence and good problem-solving skills. It is for this reason, in part, that famous rock stars and artists – think Mick Jagger with eight children or Charlie Chaplin with eleven – seem to have an above-average reproductive success. Women seem to find them irresistible despite their oft-reported propensity to stray. We'll re-encounter this in Chapter 10. So, while we might not be impressed initially by our date's outward appearance, there is still the distinct possibility that we will be bowled over as soon as they open their mouth.

And Inhale . . .

Finally smell comes into the equation at this stage, at least for women. While I am still highly dubious about the existence of human pheromones (the lack of any concrete evidence in humans, as opposed to a significant body of evidence for their presence in lesser mammals, and the large cognitive element of our dating and mating behaviour makes it highly unlikely that they have been preserved in the human line) it is truly the case that women can smell genetic compatibility, something which is of key importance if we want to produce viable offspring. Specifically, women are able to smell how compatible the set of genes underpinning our immune response – the Human Leukocyte Antigen (HLA) – are with those of a potential mate.

Ideally we want someone whose genes are as different from ours as possible to give our children the most diverse and flexible immune response to disease. Luckily, because of a genetic coincidence – the same set of HLA genes underpins our immune response and aspects of our body odour – women are capable of smelling what type of HLA genes a male has. This fact is embodied in the famous T-shirt test, which has amused psychologists and TV science audiences for years. In this, a group of men are asked to wear a T-shirt for twenty-four hours without wearing any deodorant or aftershave and without showering. After the twenty-four hours are up, the T-shirts are placed in individual jars or ziplock bags and a group of unsuspecting women are asked to give them a good sniff. What you find is that the T-shirt which a woman finds the most attractive smell-wise has been worn by the man who has the set of HLA genes most distant from her own. It's all excellent fun and certainly makes for good telly, but today you can bypass all the faffing around with T-shirts and bags and just send some spit off for analysis. For there are companies who now claim to be able to assess the compatibility between you and your partner simply by analysing the relevant bit of genetic code. I think this is great and I have certainly encountered couples who have given it a whirl. But it does always leave me with two questions. Firstly, at what stage of the relationship do you spit? Is it at the very start when your eyes have locked across that crowded room and really it might be best to check for compatibility now so that no time is wasted on a doomed relationship? I am not sure that whipping out a saliva sample tube and then asking them to wait six weeks before the first date to allow the results to be in is the best first move. Or do you wait until you are happily coupled up? What do you do if you are getting on amazingly, but the result is 4 per cent compatibility? Joking aside, it is important to remember that there are so many things that go into human attraction – physical, genetic, psychological, neural and cultural – that a lack of compatibility in one area really isn't much of a problem. So, get your compatibility certificate for the living-room wall, by all means, but just remember to keep the result in perspective.

Isn't This All a Bit . . . Un-PC?

When I describe the environment of human dating and mating, there are always two questions that raise their head. The first is linked to the distinctly non-feminist nature of the story I have told. Surely in an age where women can earn their own money, we no longer need men to protect and provide, and we must bring more to the table than just our fertility, generally denoted by being young and in good health? This certainly is the case, but two things need to be borne in mind. The first is that evolution is cripplingly slow in most cases. For something as ancient and hardwired as our mate-selection behaviour to change, we need a bit more time than the fifty or sixty years since the advent of the pill and cultural and political changes which meant that women had more freedom to control their reproduction, venture outside the home and make their own careers. Secondly, this change in mating behaviour is only likely to occur when the vast majority of the female members of our species have had the chance to be impacted by the doctrines of female empowerment, and sadly this is simply not the case at present. Globally, for the vast majority of women, feminism has not touched their world and until it does our mate preferences will not change.

The second question touches on sexuality. It is quite normal for people to point out that my story applies only to heterosexual dating. It is the case that the vast majority of mate-choice research has been done on heterosexual couples and the lack of research on other sexualities, in particular homosexuality, is surprising. But academia tends to start with the largest population and then work its way down, and as academia is shockingly slow moving, and studies require considerable replication before we can draw firm conclusions, our knowledge is decidedly heteronormative. The results that we do have paint a really rather confusing picture. For example, in some studies gay men mirror the behaviour of female heterosexuals but in others they are more closely aligned with heterosexual men. It is certainly the case that gay men place the most emphasis on visual appearance, followed by heterosexual men, heterosexual women and then lesbians. What I can say is that the argument that gay men and women must be selecting mates on different criteria than heterosexual men and women because they cannot have children, and therefore mate value must be calculated differently, is not the case. In my other research life,

I have the huge privilege to follow men as they become fathers for the first time and some of those men are gay. In the last decade or so, it has become easier to become a parent if you are not heterosexual, at least in the west, via adoption, sperm donation or surrogacy. Many gay people *are* selecting their mate in part to be a good, stick-around parent for their child, regardless of whether that potential partner is going to be the biological parent of the child or not. In this case fertility, health, resources, protection and commitment are all factors that need to be considered. It just might not be quite so clear cut who displays which attribute.

BFF* (*best friends for ever)

While you can't choose your family (or can you? We will return to the subjects of 'chosen families' later in the book), you do choose your romantic partner, hence the need to employ all your sensory and cognitive resources in making your choice. But what of that other key relationship in your life for which you have absolute choice – your friends? How do we go about deciding who is the friend for us? Who do we find attractive? One of the first studies I carried out at Oxford was an analysis of how people chose their romantic partners and their best friends. I asked the participants to what extent they shared the following attributes with their lover and their best friend – levels of physical attractiveness, creativity, intelligence, education, financial potential, sense of humour, outgoingness, athleticism, dependability, cooperativeness, social connections, kindness and optimism. What was important in each case? I was trying to understand whether there was a 'friendship market' as there was a 'dating market'. From an evolutionary standpoint, our friends do contribute to our chance of survival

– we touched on this in Chapter 1 – so we should take some care in choosing them, and it is very probable that some people are more valuable as friends than others.

We are very similar personalities, somewhat optimistic, happy to joke around, don't necessarily take every experience seriously. When we first met, it wasn't living by the seat of our pants necessarily but going with whatever happened, going with the flow and not worrying about the outside world. I may not see them for months at a time but when we get together it is as if I have spent every day with them. **Robert**

What I found surprised me and challenged the idea that our friends can never be as close to us as our lover. For women, their same-sex best friend was someone with whom they shared *more* emotional intimacy than their male lover. For men, their same-sex best friend represented ease of interaction and a sense of humour – someone you could truly relax with. Further, for both sexes, they shared more in common with their best friend – that is, they were more similar to them than they were to their lover. These results perhaps point to the inherent tension that exists at the centre of all heterosexual romantic relationships. You have two people whose sex means they approach relationships slightly differently and because of their different parenting roles are bringing different 'resources' to the relationship. I mentioned in Chapter 1 that cross-sex cooperation is cognitively the costliest of all cooperation because of the need to trade unequal currencies, and because you are having to mentalise – or mind read – a brain that probably operates in a distinctly different way from yours. With best friends, particularly of the same sex, these tensions are not there, meaning that you can truly relax and reveal your authentic self. What physical or intellectual indicators we use to assess how like us a potential friend is are still unclear, but as we practise assortative mating in a romantic context – the members of long-term romantic relationships tend to have similar market values – so we practise it with our best friends. However, in this case it is not potential reproductive success that is the driver, but finding someone who ticks the same boxes as you do and approaches life from the same perspective. We'll return to friendship as a meeting of minds in Chapter 3.

Shared interest. People who fill you up. Full of energy. Those charismatic people your personality type is naturally drawn to. They fit nicely with my personality. But then as you scratch the surface and they become deeper, you find out you share the same values and foundations which means you can have conversations at all those different levels. They share my drive, creativity, fun loving. **Matt**

Mind-blowing Attraction

So, you hold your baby for the first time, your eyes lock with a potential lover across the train carriage or a new friend across the university canteen. The brain takes in information from the senses, the algorithm runs its course and a big and resounding 'yes' is the result. Oxytocin and dopamine are released, and the game is on. But what do we see in the brain? The early stages of attraction are largely unconscious and occur in an area of the brain called the limbic system. This is the area at the core of your brain where your emotions sit. Initially, areas within the limbic system which are rich in dopamine and oxytocin receptors are stimulated. These include the nucleus accumbens, a round structure which is so crammed with oxytocin and dopamine receptors it lights up like a Christmas tree. This activation enables you to have the motivation and confidence to make that first contact. But as time passes and the attraction deepens, the signal does not remain static, but slowly moves into another area of the limbic system known as the head of the caudate. This shift is key because it is at this stage that your attraction, and possibly lust, begins to evolve from one which is largely unconscious and based in reward and novelty, to something deeper and more conscious. You are on your way to love. This is because the head of the caudate has numerous connections to the many-folded outer layer of your brain – known as the neocortex – where your conscious brain sits. And this link between the unconscious and conscious enables that key feature of human love to emerge, the ability to conceive and experience love at both the conscious and unconscious level. So yes, to have passion, sexual desire or the drive to help a vulnerable infant, but also to experience companionship, trust, empathy and cooperation. Ruth Feldman, the Israeli neuroscientist who I would argue has done more for the study of human love than anyone, suggests that this shift from unconscious to conscious love allows relationships to build a foundation based upon shared goals, shared environment and reciprocity.

And it would appear that, despite being such a complex phenomenon and there being a whole spectrum of different types of human love, the pattern of brain activation seen in the brains of those in deep romantic love is unique to this type of love, akin to a fingerprint. Neurologists Andreas Bartels and Semir Zeki were the first team, way back in 2000, to explore in detail the brain activity of romantic love. They recruited a group of heterosexual people, eleven of whom were female. They asked each to write a short essay describing their relationship and to fill in a measure of passionate love which gave an objective measure of the intensity of their love. Following this, those whose results showed them to be ' truly, deeply and passionately in love' – there's a description of addiction if ever I heard one – were selected to proceed to the next stage. These seventeen were placed in a fMRI scanner, a machine which allows us to see brain activation in real time, and asked to view pictures of their lover and three of their closest friends. These friends were of the same sex as their lover and the friendship had endured for a similar time period as the romantic relationship to ensure that like was being compared with like. On analysing the scans, they found a distinct difference between the activations caused by viewing a picture of a friend and that seen for the lover for both men and women – there is no sex difference in the neural fingerprint of romantic love. In romantic love, the head of the caudate and the putamen, both areas in the unconscious limbic brain, were highlighted, as were areas of the prefrontal cortex where key conscious social behaviours such as trust and empathy find their home. But as key as these activations were the *deactivations*, areas which showed less activity in the romantic love brain than the friendship love brain. These areas included the amygdala, showing a reduction in fear and risk detection, and the medial prefrontal cortex, which has a role in mentalising, the ability to understand or predict the intentions of others which we encountered in Chapter 1. These results are striking because they provide evidence that the belief that 'love is blind' is not just an old wives' tale. It is genuinely the case that when we are in love we become less adept at assessing the potential risks of being in the relationship and our ability to correctly understand someone's intention is clouded, opening us up to emotional and physical harm. It is truly the case that when it comes to spotting a wrong'un, it might be better to listen to your friends than rely on your own judgement.

The Passion of Parenting

My most powerful love is with my own child. I have fierce admiration for her small struggles to become. And now as I experience the adult child as full-hearted, compassionate, intuitive, using her intelligence wisely in the world. It is love transmitted onwards into the future. **Nicola**

And what of loves other than the romantic? Those with family, friends and children? In 2004 Bartels and Zeki returned to their field of interest, but this time it was mothers who were placed in the scanner. They viewed pictures of their children, who ranged in age from nine months to six years. There was considerable overlap between romantic and maternal love in both the conscious and unconscious brain patterns – the caudate and prefrontal cortex were again active while the amygdala was deactivated. This is unsurprising, given the close evolutionary relationships between romantic and maternal love; both are focused on the successful perpetuation of the species. But differences did exist. For mothers, an area within the unconscious brain which has close ties to maternal behaviour, the PAG, was active, while for lovers the hippocampus and hypothalamus lit up, possibly highlighting the significant role for memory in maintaining long-term romantic attachments and the absolute requirement for the sex hormones – released by the hypothalamus – which generate lust.

It is fair to say that today, even as fathers become more involved in the care of their children, studies on parental love tend to overwhelmingly focus on maternal love. However, in recent years, as those of us who study the science of fatherhood have gathered a head of steam, it has become clear that unlike the fingerprint for romantic love, where the sexes are the same, there is a difference between the fingerprint for paternal love and that for maternal love. In 2012, fifteen pairs of heterosexual parents of six-month-old babies volunteered to be placed in a fMRI scanner and have their brain activity assessed while they watched videos of their children playing. Israeli psychologist Shir Atzil wanted to explore whether differences existed between maternal and paternal love. Both mums and dads showed activity in the areas of the brain linked to empathy and mentalising, essential skills if you are to care for someone, for it allows you to feel what they feel, respond appropriately and anticipate what they may need next. This ability is fundamental to a secure attachment between parent and child, and the activity patterns in the brains of both parents showed that both mum and dad have the neural capability to build a strong attachment to their child. We'll meet the concept of attachment again in Chapter 3.

What do my kids come to me for rather than my wife? Maybe it's around the things I enjoy, the things I like doing with them, so if they want a bike ride they will come to me. Sarah tends to do more creative things with them; I come home and there are paints and crayons [and I think], Ooh, looks like hard work! **John,** dad to Joseph (four) and Leo (two)

But in other areas of the brain, there was a distinct difference between the sexes. In the brains of the mothers the evolutionarily ancient limbic system was the most active. The fact that Shir saw more activity here in mothers than fathers may reflect the key characteristics of mothering – giving affection and nurturing. In contrast, in the father's brain it was the neocortex that was set alight – the outer, deeply riven surface of the brain. In particular, the areas associated with social cognition – responsible for enabling someone to handle complex thoughts and tasks and make plans. This may reflect the special responsibility a father takes above and beyond that carried out by a mum for teaching and encouraging his child to strive towards independence. I've written more about this in my book *The Life of Dad*. But for someone interested in the evolution of love, the fact that the peak in activation for mum was in an ancient bit of the brain while that for dad was in the relatively young neocortex reflects the different evolutionary time points of the emergence of maternal and paternal love. Mothering is as old as time, present in the earliest reptiles, whereas we know that human fatherhood is only 500,000 years old at best, meaning some aspects of it are hardwired into the newest area of the brain. Further, to avoid redundancy, evolution has shaped the brains of mothers and fathers to be motivated by their love to focus on different aspects of their child's needs to ensure that, together, they meet all her developmental needs.

I think if mother is more protective and concerned in general for the baby, I think the father is more experimental, pushy. I was trying to do stuff like trying to get her to stand up a lot earlier than I think Sarah would have done, to see if she could do it. It is something I have noticed between us, I am more likely to push the boundaries a little bit. **Dan,** dad to Jasmine (six months)

Blood is Thicker than Water

The love I have for my friends is the same type of affection and emotion I have for my family. It is on the same spectrum but on a different scale. I would do anything for [my best friend] Nick, but I would also murder Nick for my son! It is a sliding scale, they are on the spectrum but there are degrees of separation. It is the same type of feeling. That high. It feels the same but is dialled up with family. **Joe**

It is fair to say that our knowledge of the neural activity underpinning romantic and parental love is more advanced than the love that binds us to our wider family or our friends, perhaps a reflection of the relative importance we place on these relationships. But in a first step into this arena my colleague Rafael Wlodorski explored how our brains act when we are interacting with our friends as opposed to our family – or kin, in technical parlance. In a paper snappily entitled 'When BOLD[1] is thicker than water: processing social information about kin and friends at different levels of the social network', Rafael used fMRI to compare family and friend relationships which were at similar levels of intensity, i.e. they sat at the same level of the social network structure that we encountered in Chapter 1. Twenty-five female participants were recruited and asked to provide details of three family and three friends – excluding a romantic partner – at each level of the social network. Ultimately ten individuals – five friends and five family members – were selected for each participant. Once in the scanner, they were tasked with answering questions about each individual which were designed to tax their brains – for

---

1   BOLD stands for blood oxygen level dependent imaging.

example, 'I think X is not interested in abstract ideas' or 'I think X generally feels other's emotions'. What he found was that when it came to the cognitive elements of love – that's the conscious bits seated in the outer layers of the brain – answering questions about friends generated much more activity, indicating more cognitive effort, than answering questions about family, when comparing like with like. In particular, the areas of the brain linked to mentalising – understanding how others might think or act – were particularly engaged with friends. The participants were having to think harder to answer the questions about their friends – and by association understand their friends – than their kin. Why is this? It is because when it comes to kin there is a shortcut to trust *because* you are related to them. It is the kinship premium which means that if you were to ask a favour of a family member, or vice versa, you are more likely to carry out or accept the favour than if they are a friend of the same degree of intimacy, even if they sit in the crowded 150 layer and you meet very intermittently at family dos. And why does this work? There are many factors, but it is partly because by helping them you are helping yourself due to a degree of shared genetic material – this is kin selection – but also because the family is a very interrelated network through which news travels quickly. If you were refused help, or refused to help, granny would quickly hear about it and come down on you like a ton of bricks!

As well as highlighting the key emotions and cognitions which underpin both long-term romantic and maternal love, the areas identified by Bartels and Zeki in their ground-breaking research were rich in both oxytocin and dopamine receptors, indicating that a role for these neurochemicals remains once attraction has become love. However, a short paragraph in their 2000 paper on romantic love hints at something more, a different chemical mechanism which might come into play when we are in it for the long term. They commented, simply in passing, that the pattern of activations they had seen on the fMRI screen mimicked, in part, those seen when subjects are given drugs which mirror the actions of cocaine and heroin. Specifically, they stated that the activations of the anterior cingulated cortex, part of the prefrontal cortex, the caudate and the putamen were identical, leading them to conclude that there was 'a potentially close neural link between romantic love and euphoric states'. Is it possible that something beyond dopamine and oxytocin, something whose effects mirrored those of an opiate, was in the mix?

Euphoric Endorphins

I began my life as a primatologist and one of the pressing debates of the early 1990s was the role of grooming in some primate species. I am sure we have all seen the care with which monkeys will work their way through the hair of a fellow group member while the recipient sits, eyes half-closed in euphoric bliss. For many years this was believed to be a purely hygiene-related activity. But when I began my degree in 1995, the consensus was shifting towards the possibility that grooming was not a utilitarian activity but one which had a key role in underpinning the complex social networks that exist in primate groups. For if you analysed both when and who was groomed, there was a clear association between grooming and periods of reconciliation and relationship maintenance, currying favour with those high up in the hierarchy or those from whom you would like to receive sexual favour or support. It certainly wasn't the case that those with the lowest parasite load had received the most grooming. So, if two former allies had clashed, their bout of aggression would be followed by a long bout of mutual grooming, and being the alpha male or female meant that you often received grooming from, but never gave grooming to, several subordinates at a time. However, it was only by accident that scientists discovered what it was about grooming that made it such an effective and blissful social lubricant.

The first study exploring the role of beta-endorphin in monkeys was intended to elucidate its role in male sexual behaviour. Primatologist Paul Meller and his team hoped that by administering both an endorphin agonist, in this case morphine, and an endorphin antagonist, naloxone, they could explore the impact of endorphins on a male's motivation to copulate. Agonists are chemicals which mimic the action of the neurochemical while antagonists prevent the neurochemical having any impact on the subject. They 'antagonise' its effect – meaning, in this case, no euphoric high. Unfortunately for Paul and his team, their results were decidedly mixed and didn't help clarify the question of male sexual behaviour, but what they *did* show was that monkeys who received the antagonist, meaning that if endorphins were released they had no impact in the brain, were almost obsessive in their desire to groom and receive grooming from troop mates, whereas those who had had morphine had no interest in being groomed. In the first instance the monkeys were desperate to receive the opiate high triggered by a good bout of grooming, while in the latter case their needs had already been fulfilled by morphine so there was no incentive to seek out a willing grooming partner. While Paul may have been disappointed by his inability to advance our knowledge of male sexual behaviour, he and his team had uncovered the secret behind the desire for monkeys to groom; they received a euphoria-inducing endorphin hit.

Since then, particularly through the work of my research group at Oxford, the evidence for beta-endorphin being the glue of long-term human relationships has steadily grown. It has been a tough chemical to study because, unlike oxytocin, you cannot simply take some saliva or blood to check whether it has been released following an interaction. Beta-endorphin does not cross the blood–brain barrier, meaning that the only way to access it is either to subject your participants to a highly invasive and risky cerebrospinal tap – not something that encourages many volunteers – or undertake expensive rounds of PET scanning which allows the actual release of beta-endorphin in the brain to be measured. We have been very lucky as a group to collaborate with a team at Aalto University in Finland, led by neuroscientist Lauri Nummenmaa, who are experts in PET scanning and the work we have done there has shown that beta-endorphin does indeed have a role in social interactions in humans, even that there is a link between its effects and the nature of your attachment, something we will explore further in the next chapter. In one of the first studies we published, in 2016, we explored the role for beta-endorphin in touch in long-term romantic relationships. We recruited eighteen male participants in their twenties who agreed to be placed in a PET scanner and scanned while their romantic partner – they had all been in romantic relationships for between one and four-and-a-half years – stroked them all over their body, avoiding the genitals. The endorphin activity in their brains was compared to that during a baseline scan, which involved simply lying in the scanner for an equivalent period of time as the touch task, and when being stroked by a male stranger – agreed to be the most uncomfortable form of touch for heterosexual males. The logic for using all males as our participants was that if we saw significant endorphin change in the less emotionally driven sex while being stroked by their partner then we could be pretty confident in our findings. The results confirmed a role for beta-endorphin in human touch when being touched by a romantic partner. We had the first piece of our evidence.

Spitting for Science

Being a scientist can sometimes be a mucky business, and in 2014 I and my colleagues at Oxford, Eiluned Pearce and Rafael Wlodorski, decided that what we needed to do was spend a lot of our time visiting the science festivals of the UK and getting covered in spit. There was a serious point to our activity. We wanted to explore the key roles for dopamine, beta-endorphin and oxytocin in both our closest relationships – in this case, those we have with our lover – and within the wider community. We wanted to do this because while previous studies had confirmed that these three neurochemicals were key to our relationships, and we understood some of their genetic underpinnings, no one had explored the genetics of all three together, nor had their role in the creation and maintenance of the wider social network been looked at. Hence the spit.

We decided to look at three different areas of social behaviour. The first was social disposition – how sociable you are in general; the second, dyadic relationships[2] – in this case, the relationship with your lover; and the third was social network – how embedded you are in your community. We asked over 1000 people to 'spit for science' and gathered data from them with a range of questionnaires and activities. These included the Reading the Mind

---

2 A dyadic relationship is one where there are two people involved. So a romantic couple, a parent and child, two friends, etc.

in the Eyes Test, which assesses how empathetic you are, and the Inclusion of Other in Self Scale, which is a visual representation of how embedded in your community you think you are. What our results showed was that the strongest relationships for social disposition, including your attachment style and your empathetic skill, were with the genes associated with beta-endorphin. Where romantic relationships were concerned, the genes associated with oxytocin and beta-endorphin were the most influential; and for measures linked to your social network, such as how many friends you had and how engaged you were with the local community, the stand-out genes were related to dopamine. As the hormone of vigour, dopamine seemed to be driving our enthusiasm to get involved in the world beyond our front door. But what was key for us was that yet again we had proof that beta-endorphin underpinned our long-term partnerships, with a particular impact upon the attachment and empathy which are the foundational elements of all long-term close relationships.

In slightly less invasive (and disgusting) and considerably cheaper studies, we have also shown that beta-endorphin is released when we laugh, when we touch, when we dance and sing and when we exercise. Further, if we do all these in synchrony we get a ramped-up endorphin hit; its volumes and effects are increased several-fold. All of these results were obtained as a result of exploiting the reason beta-endorphin evolved in the first instance – to deal with pain.

A Natural High

Beta-endorphin is your body's natural painkiller, but over time it has been co-opted for use within the social sphere. This may be because of its role in neurochemical reward and the reduction of stress, meaning it is ideal for underpinning our, at times, long and tricky relationships. Bearing this in mind, we conjured up a novel way of finding out whether or not it had been released, by measuring whether someone's tolerance to pain altered as a result of interacting socially or producing a behaviour, such as exercise, which we knew caused an endorphin release. This has led members of my team to follow the Oxford rowing eight, subjecting them to a pain-tolerance test – delivered via an ever-inflating blood-pressure cuff – before and after they rowed alone and as a team of eight. In the first condition the individuals tested did show an increase in their pain tolerance, indicating that, as expected, their physical exertion on a rowing machine had caused beta-endorphin to be released. However, for us the more powerful and relevant finding was that if we got them to row together in synchrony then their pain threshold shot up even further, showing that carrying out the behaviour as a synchronous team led to a several-fold increase in release. In another study, one of our PhD students attended the Edinburgh Fringe Festival and subjected some volunteers to pain-threshold tests before and after watching a comedy show (I assume she checked beforehand that it was a good one!). As predicted, following the show the audience's pain threshold had increased, showing that laughter is indeed the best medicine. Finally, my colleague Bronwyn Barr asked ninety-four willing participants to take part in her silent disco. Before they got the opportunity to dance, they were taught a series of dance moves which they then did either in synchrony, asynchrony or completely independently. Bronwyn found strong evidence for endorphin release in each case – dancing after all is exercise – *but* those who danced in synchrony received the *biggest* endorphin release and were the most socially bonded following the dance session. In an extension to her study, Bronwyn then took another group and repeated her experiment, but employing the antagonist naltrexone, which acts to block endorphin receptors. Using this is really the gold standard of proof if you are not going to the extent of investing in scanning. In this double-blind study, a third of her participants received nothing (the control group), a third received a placebo and a third received a dose of naltrexone. Following the dance session, her results showed that the naltrexone group had experienced a *decrease* in their pain tolerance, and as expected the control group's tolerance had risen due to exercise-induced endorphin release, proving that the phenomenon she had observed in her first study was as a result of endorphin release.

I'm aware of a strong group energy that's developed during singing and from the music-related stories people have brought and the personal nature of some of them that deep connections have been made that can only enhance people's sense of safety and belonging. ... the joy that singing brings in coming together in trust with others. **Rebecca**, choir member (taken from Pearce 2016)

From these studies it is quite clear that not only is betaendorphin the glue which underpins social bonding, but that if we carry out an activity that allows us to act in synchrony, the impact it has on our sociability, on our love, is multiplied. This means that not only is it capable of bonding two people but that it also has the power to bond whole groups – teams, army units, church congregations. Just think of all the organisations which use a synchronous behaviour in their activities or worship – church choirs, marching armies, chanting football fans. Those activities are not there merely by accident. They have a key organisational role, something we will return to in Chapter 9.

The Universal Love Chemical

So, what is it that makes beta-endorphin such a great candidate for underpinning long-term human love? Beta-endorphin is an astonishing chemical. It is not only the body's natural painkiller but is fundamental to the operation of many of the body's vital functions, including digestion and the regulation of the cardiovascular and renal systems. But it is within the brain that its power is really apparent. Beta-endorphin receptors are located in all the brain's key areas, including at the core of the brain within the limbic system, and on the outer surface within the neocortex. This means it has a role in the experience of the most basic of emotions, including fear and love, and in the more considered and cognitively more demanding decisions of the conscious neocortex, which include the ability to navigate our socially and technologically complex modern world and – for our purposes – fall in and remain in love. It is this wide-ranging brief that secures beta-endorphin its place as the king of bonding chemicals, because it is capable of supporting all the many subtle nuances of human relationships.

And the fact that it is released by so many behaviours that we can enjoy with anyone – laughing, touching, dancing, singing, exercising – means that beta-endorphin is capable of underpinning any form of human love. While great for the lesser mammals, oxytocin alone is not capable of underpinning human relationships. In the first instance, over time we build a tolerance to oxytocin which means that its impact on us is reduced. Secondly, it is released in any great measure only in a limited range of situations, the vast majority of which are related to reproduction – childbirth, breastfeeding and orgasm. As a consequence, it is not capable of underpinning those relationships which are decoupled from sex or childbirth which, last time I looked, are two activities which are not a factor in the vast majority of friendships. And friendships, as we know, are as critical as our relationships with our lover and our children when it comes to health and happiness.

Love is a very deep respect, understanding, attraction and that reciprocal feeling. You know you get it back. There's a bit of longing if you haven't seen them. Missing somebody. I have it for my friends in Edinburgh who I don't see very often. I have that almost physical need to see them, hug them, be around them. **Louise**

But more than this, beta-endorphin's great power is its addictive quality. It is the body's natural opiate, like heroin or morphine, so once someone has experienced an interaction that causes a release of this chemical, they will keep on returning for more. They are addicted to the wonderful feelings of warmth, closeness, euphoria and happiness that it engenders. When we interact with someone we love, we get a hit of opiate, but if we go away our levels drop, our cravings begin and we are motivated to return to the source, meaning that we are constantly drawn back to the relationship. The downside of this is that if we get dumped we go into a massive opiate withdrawal – we go cold turkey – which is why losing love is such a physically and psychologically painful experience. When we are in a relationship, we get used to existing at an elevated level of beta-endorphin which, as well as making us feel amazing, is acting as a painkiller. When we are dumped, this level plunges back to baseline and suddenly all those physical niggles, the symptoms of which were masked, make their presence felt again. Combine this with the impact that the loss of beta-endorphin, oxytocin, serotonin and dopamine – all happy chemicals – has on our mental health and you have a recipe for feeling pretty awful.

Baby Love

The evidence for beta-endorphin underpinning the bond between lovers and between members of the same team or army unit is pretty conclusive. However, it is only very recently that we have obtained conclusive proof that endorphins also lie at the heart of the bond between parent and child. In her 2016 paper published in *Brain, Behavior, and Immunity*, Adi Ulmer-Yaniv and her colleagues explored the release of beta-endorphin, oxytocin and interleukin-6 – a new chemical to add to our repertoire, which is implicated in the stress immune response – in the formation of bonds between lovers and between parents and their children. Ulmer-Yaniv argued that a combination of the body's affiliation, reward and stress systems is involved in the development of attachment in these closely bonded relationships. As we know, oxytocin lowers the inhibitions to forming relationships (affiliation), beta-endorphin provides an addictive reward and interleukin-6 represents the inevitable stress associated with forming a new bond – we all remember the uncertainty surrounding our first love. Her study involved

twenty-five heterosexual couples who were in a new relationship, 115 first-time mums and dads of four- to six-month-old
babies and twenty-five single people who acted as a control. She
took blood samples from all her subjects so she could identify
the biomarkers of the chemicals of interest and then asked those
in relationships to carry out an interaction in their couples. For
mums and dads, that was separately playing with their baby
for ten minutes, while romantic couples had to plan 'the best
day ever' to spend together. Researchers noted to what extent
the subjects displayed a range of positive bonding behaviours
– shared gaze, facial emotion, vocalisations and touch – and to
what extent they were synchronous. What they found was that
in those relationships where new bonds were being formed –
that's parent to baby and between lovers – levels of all three
chemicals were significantly higher than in the singles group.
Further, new parents had higher levels of beta-endorphin and
interleukin-6 than the new lovers, whereas the levels of oxytocin
were higher in the new-lover group. What does all this mean?
It means that we have conclusive evidence that beta-endorphin
is the key chemical underpinning the long-term bond between
baby and parent, that becoming a parent is more stressful than
falling in love (!) and that oxytocin, while significant particularly
in the short-term – hence its high levels at the start of the lovers'
relationship – does not act alone to bond parent to child or lover
to lover.

Michael Liebowitz was acting on little more than clinical observation when he wrote *The Chemistry of Love,* but the evidence is
now strong that love is as addictive as any opiate and, like heroin,
it can easily control your life. But whereas a life lived with heroin
can only end badly, a life lived with love has the potential to bring
happiness, contentment and health. And in the next chapter

we explore why I believe that the most powerful love is only to be found where an attachment – the intense relationships with friends, lovers and family which scaffold our lives – gives us a firm foundation from which we can engage with the world and find life-long satisfaction and success.

CHAPTER THREE

ATTACHMENT

● ● ● ●

Observe the scene. A doctor's waiting room. A toddler and his father are waiting to be seen, the toddler intermittently playing with toys and clambering back onto his father's knee for a hug, his reserves of cuddles seemingly requiring a top-up. An old lady with a stick walks slowly into the room and the toddler scans his father's face to see his emotional reaction. Dad's apparent calmness provides reassurance that she does not represent a danger, and all is well for the little boy. He returns to the toys. The dad's name is called from reception and, after checking that his son is fully absorbed in the demands of a shape sorter, he walks to the desk to answer the query. His son, glancing over to the now-empty waiting-room chair, suddenly cries out in distress as he realises that the security of dad is gone. His father crosses the room rapidly and scoops him up, hugging him as if his life depended on it. They return, a firm team, to the desk.

This scenario – familiar to parents the world over – makes real the concept that sits at the very centre of our most powerful relationships: the toddler's need for constant visual and physical contact; his reliance on his father to monitor the safeness of their environment and select an appropriate reaction; the almost physical, distressing pain he feels when they are parted; the security afforded by the relationship that enables him to have the confidence to explore the world. For the bond between this child and his father is what we call an attachment, the set of specific feelings and behaviours that some argue is the only true objective measure of love.

My most powerful love is for my children. I would literally do any-thing for them. The thought of losing them or their being hurt makes me ache physically. **Kim**

In this chapter I want to explain the concept of attachment; the deep, intense psychological state which underpins our most powerful loves and is, as far as psychologists are concerned, a valid attempt to provide an objective measure of love. We'll explore where attachment began with the pioneering work of John Bowlby and Mary Ainsworth, which focused on the critical, survival-based attachment which exists between a mother and her child, and understand how the nature of the attachment between child and parent is the foundational attachment upon which all others are built. We will then go on to see how this concept has expanded to include reciprocal relationships, such as those we might have with a lover, where both members ben-efit from the security attachment affords, enabling them to go out into the world and be the best version of themselves. We'll dive into the age-old debate of nature versus nurture to explore what underpins our individual attachment styles and we will encounter the one concept that I believe most reflects the critical importance of love to our survival and happiness: biobehavioural synchrony.

The Story of Attachment

Attachment relationships are a rare thing in your life. First identified by psychiatrist John Bowlby in the first half of the twentieth century and then elaborated by developmental psychologist Mary Ainsworth in the 1970s, they are defined by an intense emotional bond which endures over time and distance. Attachment is focused on a specific person to whom we crave proximity, although this desire does not have to be reciprocated. Bowlby's first foray into this phenomenon began with an exploration of the attachment between mother and child. When he first began his research, the dominant belief was that a child craved proximity to her mother simply because she was the source of food. But Bowlby recognised that, when separated from their mother, children displayed considerable distress which could not be remedied by the presence of another carer, even if food was also in the offing. There seemed to be something key to the bond with the mother. Further, through his work as a child psychiatrist he recognised that children who appeared to have insecure attachments to their mothers experienced considerable problems with emotional and behavioural development. Bowlby recognised that the attachment figure, through their sensitive care and nurture, provided the child with a sense of protection, security and comfort which increased their chances of healthy development and survival. It was the relationship that was the foundation for life.

Since Bowlby's time, the concept of attachment has spread beyond the mother–child relationship to encompass the father–child relationship, that between romantic lovers, close friendships and even, as we will consider in the next chapter, our pets. They are special precisely because they have such an influence on our psychology and behaviour and it is no coincidence that it is within our attachment relationships that we experience the most profound and powerful love. These relationships are those which will shape our life, including our development as a child and our health and happiness as an adult. As such, they are party to the most powerful biological, psychological and cultural influences to ensure we seek them out and maintain them. But we do not all form attachments in the same way and it is this difference, and how it influences our experience and behaviour when we are in love, which we will turn to next.

The Attachment Spectrum

The most powerful love I have experienced is my love for my boyfriend. Whilst my family love me, it's always been there. With him, feeling that love grow and being in complete awe/happiness with him, almost being overwhelmed by how much I now love him . . . **Tasmin**

The nature of an individual's attachment profile is defined based upon their position on two distinct dimensions; avoidance and anxiety. That is the extent to which they avoid relationships and the extent to which they experience anxiety in the relationship, particularly anxiety related to abandonment. How these actions and beliefs manifest themselves in the individual depends upon the nature of the attachment. If we consider romantic attachment, then attachment profiles come in four distinct types: secure, preoccupied, fearful-avoidant and dismissing-avoidant. **Secure** people are low in avoidance and low in anxiety. They are very comfortable with physical and emotional intimacy and feel empowered as an individual by their relationship. In contrast,

preoccupied people are high in anxiety but low in avoidance. They spend a considerable amount of time – hence they are preoccupied – worrying about their relationship and, in particular, whether their lover is going to leave them. They deal with this anxiety, in part, by trying to maintain as much proximity to their lover as possible – they are 'clingy', in everyday parlance. Those who are high in anxiety and high in avoidance are **fearful-avoidant**. Like a preoccupied person, a lot of their headspace is given over to worrying about abandonment, but they deal with this worry by avoiding relationships altogether, or if they are in a relationship, by avoiding intimacy, believing that by doing this they cannot get hurt. Finally, we have those who are **dismissing-avoidant**. These people are low in anxiety and high in avoidance. I often describe them as the classic island, uninterested in and unmotivated to start or maintain a romantic relationship with anyone. In some cases this can be as a consequence of a real fear of intimacy.

Why do we need to know this? As researchers it helps us provide a framework for understanding the psychologies and behaviours which we observe in our participants and allows us to try and understand empirically what the foundations of these thoughts and behaviours might be – genetic, developmental, etc. Beyond this, insecure attachments can, for some people, have a detrimental impact on their life. By understanding what underpins them we can develop interventions to help them. But possibly more importantly they are helpful to the individual concerned, particularly if they are unhappy with their relationship history. By understanding our attachment style – how we feel and behave when we are in love – we can maybe start to understand why history repeats itself in an unwelcome manner and begin to work to change this. For your attachment style is not fixed in stone. Depending on the causal underpinnings of your style, you can actively work to

overcome it. This might be as simple as picking the right partner. In my case I was definitely preoccupied when I first met my
husband, who is secure. Gradually his secure behaviour showed
me that my fears of abandonment were unfounded and over
time – we have now been together for twenty-three years – I, too,
became secure. For some people, the work is much more intense
and purposeful than this. It might be rigid self-awareness and the
ability to actively work against your natural tendency, a course
of therapy to unpick the precursors or intense behavioural and
psychological interventions led by attachment specialists. What
I will say, as I say to all the audiences who attend my talks and
who rush to identify their profile as soon as the PowerPoint slide
pops up, is that none of the styles are wrong. They all have benefits – for example, being preoccupied makes you very attentive to
maintaining your relationship, while dismissing-avoidant people
are highly independent – and if you are happy with where you sit,
then that is all that matters.

The First Attachment

*My most powerful love is the love for my children. The depth of my
love for them is immeasurable and hard to put into words. I love
them like they are a part of me and my soul.* **Seamus**

While our attachment styles in romantic relationships can hold
the key to our adult health and happiness, those we build with
our carer or carers when we are very young are perhaps the most
fundamental of all. Our first loves are the ones which will shape all
future love. These are our foundational attachments and from them
all others emerge. As with romantic attachment, we recognise four

distinct attachment styles in children: secure, avoidant, resistant and disorganised. Because it is key to ascertain the nature of this attachment at a very young age – around eighteen months is usual – unlike in the adult scenario we cannot ascertain a toddler's profile based upon questioning their thoughts and feelings. Instead, it is about observing the interactions between carer – usually the parent – and child in a specially designed experimental paradigm developed by developmental psychologist Mary Ainsworth entitled The Strange Situation. It uses a sequence of interactions between the baby, its mother and a stranger (usually a researcher) to explore the nature of the attachment between child and mum.

The test is carried out in a room with toys and a two-way mirror so that the test can be observed. In the first stage the child and mother enter the room and the child is allowed to explore the toys on their own. After a few minutes a stranger enters the room and begins to chat to the mother. The stranger then begins to interact with the child and during this time the mother leaves the room. After a few minutes the mother returns, comforts her child if need be and then both stranger and mother leave. A few minutes later, the stranger returns and interacts with the child. Finally, mum returns and reunites with her child. The role of the stranger in this scenario is not to scare the child but to confirm the child is attached to mum rather than simply any human.

The key questions the observers want to answer are to what extent the child explores the room alone, and how does the child respond when reunited with her mother? Depending upon the answers to these questions, a child can be categorised in four different ways. Children who are **securely attached** to their mum happily explore the room when mum is near, maybe involving her in their play, and, if they do get anxious, return to mum for a reassuring hug before recommencing their exploration. When their mum leaves the room they may get upset but are capable of self-soothing because they are confident of their mum's return and when she does appear they are quickly happy and soothed. These children have the confidence to explore the world because they know that their mum supports their needs and will respond appropriately, including the giving of soothing reassurance. She is their secure base.

In contrast, **avoidant** children are very independent emotionally and physically from their mum. They do not explore the room to any extent, and because they display no expectation that their mum will help or soothe them, they are unconcerned when she leaves the room. They show no preference for their mum over the stranger and when mum returns they tend to ignore or actively avoid her. **Resistant** or **ambivalent** children worry that their mum will leave them and, as such, find it difficult to be separated from her and are nervous of taking risks. To this extent they do not explore the room. In contrast, they are very wary of the stranger and deeply distressed when their mother leaves the room. However, on returning, a resistant child will show considerable anger and resentment towards their mother for leaving them, which can result in them rejecting any attempt by her to comfort them. Finally, **disorganised** children do not show a consistent pattern, sitting somewhere between avoidant and resistant behaviours. They tend to experience high levels of confusion and anxiety often because, intentionally or not, their mum scares them. Of all the four profiles, disorganised children are the ones most prone to behavioural and developmental problems, and it is often the profile exhibited by children who have been neglected.

I'm sure the above rings a bell for any parent who has dropped their kid at nursery or school for the first time. Both my daughters have been past masters at making me feel awful as they cry when I leave them, only to stop abruptly when I am out of sight (yes, I have been spying through the nursery door), but any stress of the drop-off pales into insignificance when balanced with the joy of our reunions. These are attachment behaviours in action and they permeate the parent–child bond.

Dads, Security and Challenge

While romantic attachments are the same regardless of the sex or sexuality of those involved, there are distinct differences between the attachments developed between a father and child and a mother and child. Remember from Chapter 2 the differences in brain activity between mums and dads when viewing videos of their children? Well, this difference is reflected in the nature of the attachments formed between a child and their mum or dad. The mother's attachment to her child is based on nurture and security and as a consequence it is inward-looking, exclusive to the mother and child. In contrast, while dad's attachment is also based in nurture and security, it has an added element of challenge. So rather than being inward-looking it is outward-looking, as the father turns his child's face to the world. He says, 'This is the world and I am going to teach you the knowledge and skills you need to survive and thrive in it, whatever the challenge.' So the interactions between a father and child tend to push the developmental boundaries – think of rough-and-tumble play or, when older, team sports – and tend to include elements of challenge and risk to build mental and physical resilience. However, because the child has a secure attachment to their father, they can always return to the security of dad for a hug when things get too tough.

Of course, not all dads are attracted to the idea of vigorous play or team sports, preferring to maybe read to their children or take them on nature walks. Rough-and-tumble play, in particular, is largely a western phenomenon as it is a fast track to a tight bond in a time-poor world where fathers are still comparatively absent from the home due to the nature of our work culture. Where dads have more time, the bonds are formed by much less overt behaviours, but as I say in my book *The Life of Dad*, if we unpick the time a dad spends with his child we will always find an element not only of encouragement to take the next developmental step, but of reference to the outside world.

This difference between paternal and maternal attachment is reflected in the way we assess the bond between a child and their father. Rather than creating a 'strange situation', we instead present a 'risky situation' to try to get at the element of challenge which differentiates dad's attachment from mum's. Developed by Daniel Paquette and his colleague Marc Bigras, The Risky Situation introduces the child to two apparently challenging situations. A social risk – as represented by the stranger, the same paradigm as in The Strange Situation – and a physical risk -- a set of stairs. Here the question is not only how the child explores the room and stranger, but how they tackle the stairs. Do they take on board their father's rules, do they approach them with no eye to their safety or do they actively avoid them?

In Daniel and Marc's test, children who are **securely attached** will explore the environment and interact with the stranger confidently, but with a sensible level of risk. They will obey any rules their father has set them about the stairs as they explore them. In contrast, those who are **resistant** or **ambivalent** will remain in close physical proximity to their dad, and not explore either the stranger or the stairs – any risk is too much. **Avoidant** children are reckless in their exploration of both stranger and stairs and don't obey the rules their father has set them. These children tend to take quite extreme physical risks. Finally, as with The Strange Situation, **disorganised** children show no consistent behaviour.

How to Build a Brain

The most powerful love comes from having children. It's the moment when your baby (finally) falls asleep, or smiles at you and you think – I made them. **Bill**

But why is this first attachment to our parent or carer so influential for all those that follow in life? It is because of the nature of human child development. As we know from Chapter 1, humans are born long before they should be to ensure that our large heads can still fit through our narrowed birth canals; it's that double whammy of two legs and a big brain again. As a consequence, there is still a considerable period of brain growth to go after birth to ensure we reach our full adult brain size. And one of the key areas to be subjected to an accelerated period of growth in the first two years is the prefrontal cortex; the site of your social intelligence. Research has shown us that the development of this area is particularly sensitive to the influence of the caring environment. Take, for example, the study by Israeli neuroscientists Eyal Abraham, Talma Hendler, Orna Zagoory-Sharon and Ruth Feldman, which focused on the link between both the parents' brain structure and their caring behaviour, and their child's emotional and social development. The team zeroed in on the density of neural connections in three key areas of the parental brain: the limbic system – where your emotions sit and risk is assessed; the empathetic area – allowing you to understand the emotions and feelings of others; and the mentalising area – knowing what someone might do next. All key for parenting *and* effective social interaction. Using a subject pool of twenty-five heterosexual mums and twenty-five homosexual dads – all primary carers – the team found that, regardless of whether the primary carer was a mum or dad, both the parents' behaviours and their neural structures could predict how well their child navigated the social environment of pre-school; our first real independent foray beyond the world of the family. They found that the most basic of parenting behaviours – touch, simple soothing speech and gaze – underpinned a child's ability to regulate straightforward emotions such as joy. Further, the extent to which parent and child were in biobehavioural synchrony during infancy – that's behaviours, physiological measures and bonding hormones all in synchrony (I'll explain this in more depth below) – predicted how well a child could handle their more complex emotions, such as frustration and anger, and parents who were warm and positive but used suitable levels of control and employed boundary-setting – those are the social rules being enforced – had children who were well socialised within the pre-school setting.

But beyond these clear behavioural links, there was also a striking relationship between their children's social cognition and behaviour and parents' brain structures. Again, those children whose parents had higher densities of grey and white matter in the emotional areas of their brain – that's the limbic area – were generally more positive, could regulate their simpler emotions by self-soothing and were more socially engaged. Parents who had higher densities of grey and white matter in the empathetic areas produced children who, again, were more positive, but this time employed quite complex behaviours to regulate their stronger and more negative emotions. Finally, where parents showed good levels of grey and white matter in the mentalising area of the brain, their children exhibited more socialisation – they understood and complied with adults' requests more regularly, were willing to share, and helped and comforted others. Further, and more strikingly, there was a direct link between the density of connections linking the limbic and empathetic areas of the parent's brain and their child's oxytocin level at pre-school age. It's as if their parents' brains, and the nature of the caring relationship they had with their child, were the actual physical foundation of the child's emotional and behavioural development and that includes their experience of love.

'Baby, It's All About Nurture'

It is clear from these studies that one of the most powerful influences on the nature of the attachment a child builds with their parents is their environment of development, particularly the first two years after birth. Confirmation of this conclusion has come from studies exploring the similarities and differences between the attachment relationships formed between child and parent by monozygotic and dizygotic twins. Twin studies are a valuable way of determining to what extent a behaviour is influenced by genes versus the environment, particularly when we are yet to identify which genes may underpin the behaviour, often the case when the behaviour is as complex as human parenting.

If a behaviour has a significant genetic component, then we would expect monozygotic twins, who are genetically identical, to show more similarity in their behaviour than dizygotic twins who are genetically no more related than standard siblings. In contrast, if the behaviour is largely environmentally driven, then we would expect there to be no difference between monozygotic and dizygotic twins, based on the assumption that both members of the twin couple had experienced an identical environment. In their 2004 study a team of Dutch researchers, led by Marian Bakermans-Kranenberg, were some of the first to try and understand the genetic and environmental influences on child attachment profiles using this well-regarded technique. They recruited fifty-six twin pairs, twenty-one monozygotic and thirty-five dizygotic, and asked their parents to fill in a questionnaire designed to elicit the nature of each child's attachment profile to their father. Their analysis showed that there were no discernible differences in the degree of similarity between the attachment profiles between monozygotic and dizygotic twins, which confirmed that any genetic contribution to childhood attachment profiles was minimal. Further, the largest influence was the shared environment at 59 per cent, rather than any environmental factors unique to a particular child–parent bond, such as how the interaction between the personalities of the child and parent may impact the environment, which sat at 41 per cent. These findings have subsequently been replicated by the same team in mother–infant pairs using The Strange Situation as a more robust measure of infant attachment. Again, results based on 136 twin pairs from the Netherlands and UK, fifty-seven monozygotic and eighty-one dizygotic, found that any genetic contribution was minimal and the nature of the environment, both shared and unique, was the biggest factor.

Teen Genes

Adolescence is a time of considerable change. Hormones are fluctuating, brains are de- and re-wiring, school suddenly becomes a whole lot more serious and you start to move your primary attachment relationship from your parents to your friends. This is an important shift as it begins the process of moving away from the family and becoming an autonomous, independent being. And it would appear that what influences your attachment to your parents at this stage is distinctly different from that which underpins your attachment profile as a small child. In their 2014 study, London-based psychologist Pasco Fearon and his team decided to explore the nature of attachment in 551 twin pairs, all of whom were around fifteen years in age. They asked their participants – 288 of whom were monozygotic twin pairs and 261 dizygotic twin pairs – to take part in an hour-long structured interview which explored their current interactions with their parents. Researchers scored both their verbal and non-verbal responses to questions which focused upon occasions when the teenager might call upon their attachment figure – emotional upset, illness, separation or loss – meaning that both cognitive and behavioural aspects of their attachment were considered. In sharp contrast to the work on toddlers, they found that the attachment profiles of monozygotic twins were more similar than those for dizygotic twins – 44 per cent as compared to 33 per cent – suggesting a considerable genetic input. Further analysis confirmed this, with 38 per cent and 35 per cent of the secure and insecure (preoccupied, dismissing and disorganised) profiles respectively being due to genes.

Why this difference? It may be that as the relationship between parent and child evolves, and they learn about each other's individual personalities, the parent's input becomes tailored to each individual child, which, to some extent, will be influenced by the child's genes. As a consequence, genes become influential and the impact of the shared environment – where parents simply parent each child exactly the same – fades into the background. It could be that as a child matures their concept of the attachment they have to their parent becomes more cognitively based – i.e. they reflect upon and build a coherent picture about the relationship – rather than primarily being influenced by their parent's behaviour. It is not yet clear. But what these results do show us is that attachment is a fluid concept which changes in type and cause over the life course, and that the answer is more complex than a simple ratio between environmental and genetic causes.

Attachment in the Scanner

The most powerful love I have experienced is with my husband and two sons. With my husband, it was incredibly intense initially and a very physical love, which has developed into a solid, steady, joyful, loving companionship. With my sons, it is a constant and uncondi-tional sense of awe. **Nikki**

I have studied love and attachment for over ten years, and when I first started I would have argued, based on the extensive psy-chological and behavioural evidence, that the story of minimal genetic input to attachment that we see in early childhood is one which would be reflected in our other key survival-critical attachment, the romantic relationship. But as we delve deeper and deeper into the genes which underpin the neurochemistry of love, it would appear that the story is considerably murkier than we first thought. For example, take the oxytocin receptor gene

(OXTR). This gene is highly polymorphic – meaning it comes in many different forms – and as a consequence it underpins a considerable percentage of the differences between individuals in the ways they act and feel when in love. One particular version of this gene does seem to have an influence on the nature of our attachment to our lover, but it only expresses itself if certain environmental conditions are present. If you carry this version of the OXTR gene, then you are more likely to exhibit one of the insecure attachment profiles – dismissing or avoidant – but this gene will only express itself in you – the phenotype – if you have suffered poor mental health. This is a case of differential susceptibility; the gene makes you more or less susceptible to your environment. It is likely that this sort of mechanism also explains why some people are more susceptible to the environment of their upbringing than others. I certainly have met people who have had the most neglectful childhoods but have grown up largely unscathed, which is perhaps not what you would expect. I say that they must have a cast-iron set of genes, but it is more likely that they carry genes which make them less susceptible to the influence of their environment. We'll explore this idea more in Chapter 5, when we discover the influence of our genes on individual experiences of love.

While twin studies have enabled us to understand to what extent attachment has a genetic element, they don't further our understanding of exactly which genes may play a role. However, while exploring the role for beta-endorphin in long-term human relationships, we, alongside our Finnish collaborators from Aalto University, experienced one of those serendipitous moments in science when an experiment testing one hypothesis unexpectedly gives you the answer to another. In the mid-2010s we were merrily placing people in our PET scanner as we explored the role for beta-endorphin in love. Alongside a battery of other questionnaires, we, as an afterthought, lobbed in a romantic-attachment questionnaire for participants to complete before they entered the scanner. Willing participants for PET scanning are a rare thing, so we tend to try and collect as much data as we can while we have them in our clutches. This questionnaire, the Experiences in Close Relationships measure (ECR), is a set of thirty-six questions; eighteen explore the anxiety dimension of romantic attachment and eighteen the avoidance dimension. An example of an anxiety question is '*When I show my feelings for romantic partners, I'm afraid they will not feel the same about me*', while the question for avoidance is '*I prefer not to show a partner how I feel deep down*'. The participant must indicate, on a scale of one to seven, to what extent they agree with this statement. Following completion of the questionnaires, we would place the person in the scanner. PET scanning works by reading the radio-active output released by a radioactive chemical – called a ligand – which is structurally identical to, but chemically different from, the neurochemical of choice. When we inject this ligand into the participant's bloodstream – in this case [11C] carfentanil (11C being radioactive carbon) – it attaches itself to any free receptors in the brain and we can calculate, using the size of the radioactive signal and knowing the half-life of the radioactive element – how many receptors there are. 11C carfentanil is structurally identical to beta-endorphin and latches onto the mu-opioid receptor in the same manner. The bigger the signal we get from the ligand, the more mu-opioid receptors someone has.

And what we found was that the larger the signal, indicating a larger number of receptors, the more likely they were to be secure. In contrast, those with the least numbers of receptors, particularly in areas linked to empathy and mentalising, had an increased chance of being avoidant. The density of receptors seemed to have no impact on either preoccupied or fearful-avoidant attachment styles. This link between beta-endorphin and avoidant attachment is supported by the evidence from the opiate-abuse literature. It is the case that those who abuse heroin – but not drugs which are not opiates such as ecstasy or cannabis – are more likely to be avoidantly attached, are less able to maintain long-term relationships and satiate their need for opiates externally rather than engaging in social relationships. But this link between the opioid system – the chemical of long-term love – and attachment is also striking as it suggests that attachment and love are two sides of the same coin.

I hope by now it is clear that attachment relationships, and in particular those between lovers and between parents and children – are the stuff of survival. Within them, we seem to experience a love which is more intense, more multi-faceted and more influential than those which are denied entry to the exclusive attachment club. Indeed, attachments are so fundamental to your health, wellbeing and happiness that evolution has seen fit to engage every mechanism in your body to ensure their success. For me, the concept we will explore next gets to the very heart of what it is to love. This concept is known as biobehavioural synchrony.

All-Consuming Love

The most powerful love I've experienced is for my children. It was immediate and all-consuming. I felt like the moment they were born my heart broke open and I grew another heart for each child so I could love each with all my heart. I would honestly do anything for them and I cannot imagine this world without them in it. Jess

The term biobehavioural synchrony was first coined by the Israeli neuroscientist Ruth Feldman. It has its foundations in the concept that those who are tightly bonded, who are attached, exhibit behavioural synchrony. This is probably something that we have all observed in those we know. Think of the give and take of a child and parent at play or the shared gestures, vocal pitch and linguistic signatures of romantic lovers. But if we were to look inside the body we would find that this synchrony exists not just at the behavioural level but at the physiological level too. As the lovers or parent and child interact, their blood pressure, body temperature and heart rate also come into synchrony. And if we journey to the brain, then synchrony is apparent here too. Within romantic couples, synchrony in gamma waves – the fastest of brain 'waves' and indicative of the integration of information from many brain regions – occurs in the temporal parietal regions of the brain implicated in mentalising, social comprehension and social gaze – that's maintaining eye contact. A phenomenon not seen between strangers. And it is this eye contact – combined with the degree of synchrony in non-verbal behaviour – that has the biggest impact on the extent of neural synchrony; the extent of conversation is not a factor. This distinction perhaps indicates the evolutionarily ancient nature of biobehavioural synchrony but also its lifelong applicability to the maintenance of our closest bonds, for it enables the most profound attachments to develop from the moment we emerge as pre-verbal infants.

A Meeting of Minds

In 2017 a team from Bar-Ilan University in Israel, led by neuroscientist Jonathan Levy, used magnetoencephalography (MEG) to explore whether the neural synchrony observed in closely bonded romantic couples was evident in the relationship between mothers and their children. MEG measures the changes in magnetic field caused within the brain by neural activity. Twenty-five mother–child pairs were recruited and two years prior to the MEG session, when the children were on average aged eight and a half, they were filmed while taking part in two discussions: one planning a fun day out and the other rehearsing a subject that generally led to conflict. If my relationship with my children is anything to go by, there would be many possible topics to choose from for the latter scenario: screen time, putting dirty washing in the laundry basket, the state of their bedrooms. You know the drill. Two years later, the participants were invited back to the lab and wired up to MEG machines. They were shown portions of the videos in which they starred, and portions of the videos made of the interactions between the mother–child couples who were unfamiliar to them. Results showed that when viewing videos of their own positive conversation, mum and child showed synchrony in the gamma waves produced in the temporal parietal region, but no synchrony was evident when viewing the videos of unfamiliar mother–child couples. Further, as happens within romantic couples, the synchrony in neural activity was stimulated by synchrony in the non-verbal behaviours which mum and child employed while chatting, rather than any particular aspects of their conversation. As with romantic couples, the synchrony of behaviour was mirrored in the synchronous activations of the brain.

And where neuronal activity occurs, neurochemistry follows. As with the other survival-critical relationships, the bond between father and child is also party to the impact of synchrony. In a powerful experiment, thirty-five fathers of five-month-old infants were administered with synthetic oxytocin or a placebo – both are a quick squirt up the nose – before engaging in a period of interaction with their baby. The team, led by psychologist Omri Weisman, took saliva samples from the dads at four timed points: baseline, before any administration of oxytocin or placebo (1); forty minutes after administration but before the interaction began (2); twenty minutes after the interaction began (3); and, finally, twenty minutes later (4). Saliva samples were also taken from the babies at times 2, 3 and 4. Fathers and babies playfully interacted for a period of seven minutes.

Results showed that where a father had received a dose of oxytocin, but not placebo, he was significantly more involved in play, maintained more social gaze and touched his baby more often. As we know from the last chapter, oxytocin lowers our inhibitions to social interaction and increases our empathy, so these results are to be expected. But what was incredibly powerful was that where the dad had received oxytocin, but not the placebo, the baseline oxytocin level of the baby, represented at time point 2, rose to meet that of the artificially raised level of their father at time point 3. Simply by interacting naturally – and exhibiting behavioural and in all likelihood physiological synchrony – the baby's oxytocin level had risen to bring it into synchrony with that of their dad. The rise in oxytocin level itself in the baby is unsurprising – we know kids enjoy playing with their dads – but the fact it was exactly the same as dad's is the astonishing and powerful finding. And remember this all happened in just seven minutes of interaction.

How does this happen? At present we have no idea. But what these findings show us is not only that the mechanisms of our body and brain are tightly integrated, meaning that mind and body cannot be viewed as separate entities, but, more importantly for love, that it is truly the case that love is *so* fundamental to our lives that every single mechanism is recruited to ensure that our bond is as strong as it can possibly be. Love permeates every fibre of our being. Love truly consumes us.

This chapter has been about attachment, the profound and intense survival-based bond that exists between us and our lovers, parents and children. We've learned that attachment is an attempt to provide an objective measure of deep love and is the psychological concept which impacts how we feel and behave when we are in love. We have encountered the phenomenon of biobehavioural synchrony – the one idea I would argue that brings home how vital love is to us all. We have learned that attachment, like so many other aspects of our physiology and behaviour, is the result of a mix of environmental and genetic influences and that arguably the most critical environment for our lifelong attachment pattern is that which we experience in the first couple of years of our life. But we know that it is not only those with whom we build romantic and parental relationships who have an influence upon our health and quality of life. In the next chapter, I want to explore a love that we often underestimate but that is arguably as powerful an influence upon us as any we have with our family or lover. It is the love we have for and the attachments we have with our friends. And those friends can come in all shapes and sizes. Some even have paws.

CHAPTER FOUR
UNDERESTIMATED

● ● ● ●

I think my most intense love relationships are with friends and they are a decision, decision loves. I've had a lot of romantic relationships but I never felt that 'waah'. My experience of love has been in a friendship environment and I have had a lot of friends. **James**

Within the west, we are inclined to conceive of a hierarchy of love. The top position tends to be occupied by parental love and, because of a lack of understanding of the powerful attachment between father and child and the role dads play in their children's lives, this is usually embodied in the love between mother and child. I've written elsewhere about why this is simply not the case – dad love should be up there too. Running a close second is romantic love, and with the overwhelming focus we have on finding and keeping love – think of the proliferation of dating apps and self-help guides, the storylines of our books, films and soaps and the £12 billion global wedding industry whose members compete to sell you the idea that there is such a thing as a perfect day – you would be forgiven for thinking finding your 'soulmate' is the only true point of existence; we fail at this and we live only half a life. Following this we have the family – siblings, parents, aunts, uncles, grandparents – and maybe even the extended family. After all these, the next category comes a rather distant fourth – our friends. It is fair to say that when considering love we neglect our friendships.

In this chapter I want to consider the love we underestimate, that we have with our friends. In Chapter 2 we encountered what attracts us to our friends, but in this chapter I want to explore what keeps us there and why, for many, their friends *are* their family, fulfilling all the needs which might more traditionally be met by a lover or children. We'll explore the concept of the chosen family and its importance to those, in particular, whose life choices mean they may have been excluded or rejected by their biological kin, and we will look to the future and whether the rise of AI means that we could ever make a robot our friend. And because I am in charge, we will spend some time looking at the love we have for, and the attachments we build to, man's (and woman's) best friend: the domestic dog.

In the first instance, I want you to cut out all the noise around the primacy of romantic or parental love and look at your friendships, and their value to you, in a different way. Because our society has profoundly changed in the last fifty years. As a result, it is no longer the case that you have to be 'coupled up' to fit society's norms, to have children or, as a woman, to make sure you are economically provided for. Within the West, although not necessarily elsewhere, romantic love is a choice rather than a necessity. If children are not your thing then you can also dispense with parental love. But you discard the love you have for your friends at your peril, because these are the people who make up the vast majority of the social network whose size and quality – as we know from Chapter 1 – are the greatest factors in your health, happiness and life satisfaction. Further, they are the only group with whom you experience platonic love, and who you get to actively choose! There is no familial obligation or cultural pressure. As we know from the study I recounted in Chapter 2, friends play a key role in our adult lives, providing unmet depths of intimacy for women, and relaxation and humour for men. They are the people we turn to when we really need to open up and just be ourselves. When you are a child and enter pre-school for the first time, your world opens up dramatically and you get to instigate friendships yourself, making your own choices rather than being forced to play with your mum's best friend's kid while your mum chats over a coffee. And when you are at secondary school your friends become your primary attachment figures, the people you turn to to influence your behaviour and thoughts as you establish your autonomous identity, and the welcome source of the vital secure base once afforded by your parents. In adulthood, friendships might ebb and flow as you reach life stages at different points, but they will remain a source of comfort, advice, fun and freedom. They may even, as we will see later, become your family.

There are two emotions that are the most exciting when you are with your friends. You feel accepted and safe and understood. The sense of familiarity, knowing someone and not expecting them to be someone else and not being surprised at seeing sides of them that might not be the best. You have seen them at all stages. You know what to expect. That sense of security. **Kelly**

Friends bring a sense of freedom and relaxation to our lives. This may be in part because we are likely to pick friends who are similar to ourselves – the concept of homophily. So we tend to select friends with whom we share our sex, ethnicity and age, our behaviour, personality and degree of altruism. Indeed, my colleague Robin Dunbar, who has spent many decades exploring the dynamics of human relationships, has identified seven pillars of friendship; the more pillars you share, the stronger the friendship and greater the love. These are: language, where you grew up, educational history, hobbies or interests, musical taste, sense of humour and your world-view. It is argued that such homophily has a psychological and evolutionary benefit. In the first instance, choosing friends who are like us reinforces our views and beliefs, making us feel more confident in our identity. But it also allows us to cut down on the precious energy – known as the cognitive load – we devote to trying to predict what they may think or do next because, due to the similarity, they are most likely to do or think exactly what we would do or think in the same circumstances. And for the first time, the idea that friends think alike has some solid evidence from the scanning room.

In 2018 psychologists Carolyn Parkinson, Adam Kleinbaum and Thalia Wheatley set out to explore whether the homophily we exhibit with our closest friends was reflected in a similarity in the way we perceive, interpret and respond to the world. Was the ease of relationship we find with our friends partly down to the fact that we just think the same way? They recruited 279 students – the entire cohort from one year of a graduate programme. They asked the students to complete a questionnaire listing everyone on the programme who they would deem to be a friend. They then set about, with the welcome help of a computer, creating a social network for the class, illustrating every link, or tie, between the students. Their prediction was that the closer two people were to each other in the network, indicating a stronger tie, the more similar their neural response would be. Placing 279 students in an fMRI scanner would have been prohibitively expensive, so a subset of forty-two students was used for the scanning study. Once in the scanner, everyone watched the same set of videos in the same order. Videos were chosen to cover a range of topics and to be sufficiently gripping so as not to encourage the mind to wander. And what the researchers saw confirmed their hunch: that homophily extends beyond hobbies, ethnicity, age or sex to include our brains. The signals seen in the brains of friends – both in the unconscious and conscious brain – were more similar than those between people who were more distant in the network. And to test their model they viewed paired sets of neural activity scans, and just by establishing how similar or different these scans were, they could predict how close the two people were in the network. Now *that* is a concrete finding.[1]

The question remains: do we become friends with those who think the way we do, or by being friends do we come to perceive, interpret and respond to the world in a similar way simply by being together? Because this study only allowed us to glimpse a snapshot in time, the answer is unclear. We would have to follow a friendship from its very moment of creation. But, as with all things human, it is probably a bit of both – a degree of homophily and the influence a friendship brings to bear on shaping our behaviour and psychology.

---

1   Interestingly this finding has just been replicated in a study published in *The Proceedings of the American Association of Science* (Hyon et al 2020) which showed a similar phenomenon between those who were most closely bonded in a village. Again, members of this community showed an increasing overlap in resting brain activity the firmer their friendship was.

The Future
is Single

There is a weird idea that you are supposed to get everything from your romantic relationship but I realise the huge amount of love I have in my life. I do miss having a very specific romantic connection but I feel the love of my friends. Living in a house share with real friends makes me realise that a lot of what I thought I wanted from a relationship was a close, daily friendship really and I get a lot of that from my house share. **Margaret**

Why do I argue that neglecting your friends places you at considerable risk? Because for a significant number of people their friends fulfil the role of a romantic partner, a child and even a whole family in their lives. They *are* your survival-critical relationships. Recent data from the 2015 US census has predicted that 6 per cent of the current adult US population will remain single for their entire life. Now obviously some of these will have children – a particular interest of mine is the rise of the platonic co-parenting relationship – but for the majority this means no lover and no kids. In this case, two groups of people become the key attachment figures who will underscore your health and happiness as an adult. Your siblings – if you have them and get on with them – and your best friend or friends. Research on the power of the attachment between friends is only in its early days, but in her 2017 study looking at the role of best friends and siblings in the lives of female singles, New-York-based psychologist Claudia Brumbaugh found that best friends played a crucial role both because of our close similarity to them and our freedom to choose them. The larger a single person's friendship network, the less likely they were to exhibit an avoidant attachment style. And it would appear that our friends may know us at least as well as we know ourselves. In their study of the brain activity associated with considering our own personality – self-referential thinking – and the brain activity of a peer considering our personality, the brain activity is strikingly similar. Psychologists Robert Chavez and Dylan Wagner recruited a small but tightly knit network of eleven students, five of whom were female. Having completed the usual battery of questionnaires, each participant was placed in the scanner and directed to think about their own personality and that of each of their ten peers. When scans were compared, the brain pattern seen when an individual – let's call them A – reflected on their own personality matched that of the pattern seen in the scans for their ten peers when they were also thinking of A's personality, but *not* when they thought about the personalities of the other members in the group, say B or C. This study shows us not only that, again, friends show synchronous activity when attending to the same task but that our friends definitely know us at least as well as we know ourselves.

Friendship, Love and Prosecco

Nick is the quintessential best friend. We have known each other since school. We have had many many experiences and memories together and he is someone I can always turn to for a chat . . . someone to lift me up, to brighten me up. He is one of the people I am my truest self to. The closer the guys get, the more shit they give each other, and with me and him nothing is off the cards, everything is up for debate. **Matt**

Claudia's study focused on the friendships amongst female participants, and it is the case that there is a sex difference in both the number and nature of the relationships we build with friends and, maybe, in the love we experience with them. Repeated studies have shown that men's friendship groups are less tightly bonded and are less emotionally close than those between women. And while it is very likely that a woman can identify a best friend or two, this tends to be anathema to a lot of men. While men like to carry out activities with their friends, women tend to prefer the opportunity to share intimate chats. Why do we see this difference? It might all lie in the impact our neurochemistry has on us.

In 2018 a group of Chinese researchers, led by Xiaole Ma, reported in the journal *NeuroImage* the findings of a study that aimed to explore the different impact the administration of synthetic oxytocin would have upon the experience of emotional sharing in men and women. They recruited 128 pairs of same-sex close friends and, having placed one in an fMRI scanner and another in an adjoining room, showed both a set of images of people, landscapes and animals selected to represent three points on the emotion spectrum: positive, negative and neutral. Some of the pairs had been administered with a nasal spray of oxytocin, while others had been given a placebo. Having completed the task with their friend, they then completed it with a stranger and then alone. What Xiaole and her team found was that where women were sharing the experience of watching the images with a friend, and had been given oxytocin, their experience was significantly more positive than if they had done so with a stranger or alone. This seemed to have been underpinned by a reduction in the activity of the amygdala – the site of our negative emotions, including fear and anxiety – and an increase of activity in the reward centres of the brain, which was, in all likelihood, reflecting the release of dopamine which accompanies any increase in oxytocin. This is the fingerprint of unconscious love. However, the same could not be true of men. In this case, activity in the amygdala *increased*. It would appear that women gain a considerable benefit from sharing emotional experiences with their female friends – reduced fear and anxiety and an increased positive mood – which might explain why we relish the opportunity to catch up over a glass of prosecco and why our conversations tend to focus on the emotional and intimate, whereas men prefer to avoid these occasions at all costs and head for the football field or meet in a large group of friends instead. Can we say, then, that for women there is such a thing as friendship love but not for men? I think this is very unlikely to be the case, but until we place a man in a scanner and recreate the relationship he has with his team of mates in the lab the jury is still out.

I would say I love my friends. It is different because it is a love that is chosen and quite special. People talk about unconditional love but I think there is something special about conditional love because you are always opting into it. Conditional love is an obligation but there is something special about day by day by day choosing to stay in those relationships. It's that preciousness that builds and deepens over time. Even with their irritating habits, their imperfections just make them realer and therefore dearer. **June**

Perhaps it is unsurprising that in our hierarchy of love it is generally parental and romantic love that take the top prizes. They are, after all, the source of our genetic survival. And our extended family come third because we share a proportion of their genes. By cooperating with them – loving them – we increase their chance of survival and, in turn, gain a genetic benefit. This is known as kin selection. But for some people the importance of their friendship group extends beyond the intimacy or ease of interaction which are the benefits of friendship for most of us, to a love that equates to that we feel with our kin. For these people, their friends *are* their family.

The Family You Choose

My family are my housemates. They are the people who feel closest to me and it sort of transcends the definition and it's not as if they have specific family roles. I don't see them as siblings . . . it is just they feel close enough to have full access to what I can offer. People who I would call in my worst state and know they won't think less of me for it, which doesn't describe the relationship I have with my biological family. It is like I have chosen what I value in a family and that doesn't fit in with the values of my biological family. **Alex**

The term 'chosen family' was first coined during the 1970s and 1980s in the United States to describe the networks of friends which provided emotional support and nurture to those who had been rejected by their own family or who were excluded from legally sanctioned methods of creating a family such as marriage or parenthood. For the vast majority of incidences these were gay men and women who had been excluded by their culture and/ or disowned by their biological family and whose need for support was made all the more urgent by the arrival of HIV within their communities. These families were bound by a shared identity rather than shared blood – they were fictive kin. And while LGBTQ+ rights have come some way in the last few decades – gay couples can now marry and have children in some countries – the presence of considerable homophobia on social media and the, at times, extreme debates about transgender rights show that universal acceptance is still some way off. The need for chosen families within this community – made up of friends of all sexualities and genders – is still strong. In their 2015 paper entitled 'Family matters, but sometimes chosen family matters more', psychologists Karen Blair and Caroline Pukall explored the perception of support from biological and chosen families among a cohort of heterosexual and LGBTQ participants. They found that for those in same-sex relationships, but not heterosexual participants, the perception was of more support from their chosen family than from their biological family, and they valued the opinions of their chosen family with respect to their life decisions, including their relationships, more than the opinions of their biological kin.

And while those who pioneered this new form of 'friend' family in the 1970s have now grown old within the bosom of their chosen family, recent work among younger communities in America has shown that chosen families are as important to the lives and as vital to the security and development of young people as they have always been, particularly when it comes to one of the trickier aspects of growing up – exploring your sexuality.

I could never just talk to my family about anything. I couldn't talk to them about sexuality, about issues in my own life, because I never felt that open space. With my best friend, when my wife and I got pregnant last year, I told him the day we did the pregnancy test because I needed to tell someone and he is the person I know will want to hear everything from me. I don't feel the same level of excitement or engagement from my family when I tell them about things. **Robert**

In 2013 The Illinois Caucus for Adolescent Health (ICAH) carried out a study exploring the role for chosen and given family in discussions with adolescents about sexual identity, health and rights. They used individual interviews, online surveys and focus groups to explore the experiences of nearly 500 adolescents as they navigated this at times rocky and confusing stage of their development. The results showed that while there was a role for both family types, the chosen family was the first port of call when discussing these potentially tricky topics. In the first instance, 80.7 per cent of the participants reported that they had formed a chosen family and that, in opposition to the assumption that these would be made up of peers, 33.5 per cent reported that adult allies were part of this group. For them, having a chosen family was not simply about rejecting the adult viewpoint. When it came to discussing sex and sexuality, 73.4 per cent of young people spoke to their chosen family, compared to 52.8 per cent who spoke to their given family. But when it came to how *comfortable* they felt while doing this, 63.2 per cent were more comfortable talking to their chosen family compared to only 9.7 per cent being comfortable speaking to their given family. And while there were topics that participants wouldn't speak to their given family about – personal sexual experiences, for example – when asked what they couldn't speak to their chosen family about three-quarters said they could speak to them about anything. Why was this? For these kids, their chosen family played a distinctly different role in their lives as compared to their given family. They were the source of emotional and intellectual support and advice. In contrast, the given family played a role providing financial and educational support, with emotional support coming a distant third. And the chosen family was particularly key for those who maybe wished to explore sexual identities which were in the minority.

For transgender, gender non-conforming and genderqueer youth they were more likely to speak to their chosen family (81 per cent) than their given (59.1 per cent). And for asexual youth this shot up to 80 per cent as compared to 20 per cent.

I am a firm believer that friends are God's way of apologising for our family. I love my friends very deeply because they are my family. I do not come from a high-functioning family and even though I love my family and I accept them for who they are, it is really my friends that are my family. They know the deep dark secrets that most of the world doesn't . . . they have seen me at my absolute worst and best. I have been predominantly on my own but my girlfriends have been that emotional touchstone, that physical presence in my life when my family was largely absent. **Carol**

But how do we know that a chosen family is based on the tight bonds of love which bind most – though not all – biological families? Using the relationships between gay men and straight women, and lesbians and straight men, as her starting point, sociologist Anna Muraco explored this question as she interviewed twenty-three fictive kin couples who were based in the San Francisco Bay area. She found that all her interviewees characterised these close friends as family, and for nearly half, their relationship with their friend superseded that with any member of their biological family. Further, while a third were alienated from their biological family, for many their chosen family members weren't plugging a gap in their network but were there because of the unique and powerful love they felt for them. Further, these fictive kin often held identifiable kin roles within the chosen family. Some were maternal or paternal figures to younger friends while others were siblings and valued uncles and aunts to their friend's children. The chosen family members were also seen to fulfil the roles more often filled by biological family members, particularly with respect to financial support. Anna found that one quarter of her participants had lent or borrowed money from each other and one couple had bought a home together to join their immediate families together into one large family comprising biological and fictive kin. The interviewees also reflected on the non-judgemental, emotional support that they received, and many had plans to grow old together or even co-parent based upon a bond of platonic rather than romantic love.

And among the youth who participated in the ICAH project, chosen families were stated *explicitly* to be about survival in a challenging world. Let's hear from a couple of them:

A lot of chosen families are made as a survival mechanism. I go to them first instead of my given family because I am surviving with them – these are the people who helped you overcome something that you couldn't necessarily express. **Kami**

You expect them to have your back – it's a survival game. It's a game of being out of the 'normal' box and trying to run away as fast as possible while others try to throw you in. Your chosen family are in that game with you. **Erica**

Can there be any greater expression of love than to know that someone has your back in the game of survival?

And They Call it Puppy Love

The dog follows me everywhere, literally all day and all night. He lies beside me in bed and when I work he is lying here. He is utterly devoted to me. And I am utterly devoted to him. If I have to go to my happy place, I think about him. The relationship is intensifying to the point where we might just leave and go off together! We are definitely happier together than we are with everyone else. **James**

We build intense, loving bonds to our lovers, to our children, to our parents and to our human friends. But one of the amazing things about human love is our ability to extend it to many different people and beings. Take, for example, other species. I can remember many tearful hours as a teenager unburdening myself to my beloved dog Henry, whose death floored me for months, and today our three dogs – Bear, Sam and Scruffy – are definitely key members of my family, alongside my children and my husband. Indeed, my husband would argue they are above him in the pecking order when the hugs are being handed out and space made on the sofa. I can plainly say that I love my dogs, they are my best friends, but do they love me in return? Is the relationship a reciprocal attachment akin to that we encountered in Chapter 3, with all the behaviours, psychologies and benefits that implies?

He's amazing. He is my safety net, he is my comfort blanket. He is my best bud. We can go for walks and he is very much a similar energy to me. Happy to hang out and chill but also happy to go for a ten-mile walk. Full of personality. He is staying with my mother-in-law at the moment and I feel like I have lost a limb. I am constantly looking around for him. He is awesome. I can show you photos. I love him 100 per cent without questions. He is our firstborn. **Matt**

In 2019 a team of animal behaviourists from the UK, US, Germany and Austria were the first to use the gold standard of attachment research, Ainsworth's The Strange Situation (SSP), to attempt to identify, first, whether dogs exhibited attachment behaviours towards their owners and, second, whether they displayed the four attachment profiles seen in the infants who were more routinely the subjects of this procedure. They recruited fifty-nine medium-sized dogs and their human caregivers. There was an even split in sex for both dogs and humans. The dogs were subjected to the standard SSP and their responses analysed. What the team found was that, in the first instance, dogs showed a range of behaviours upon separation and reunion which confirmed that their caregiver was an attachment figure. And the nature of these behaviours enabled the researchers to place the dogs in one of four attachment profiles akin to those seen in human infants. So **secure** dogs actively sought contact and proximity on reunion including jumping up, making contact with their snout and paws and tail wagging to encourage contact from their owner. The secure dog would actively explore the room and toys prior to separation, and while separated they would search for their owner but didn't show overt distress. **Avoidant** dogs did not try and make contact or seek proximity during reunion. They explored the room before separation and did not show distress upon their owner leaving. **Ambivalent** dogs didn't explore the room but preferred to remain close to their owner. When they were separated, they made distressed vocalisations and searched the room, but reuniting with their owner didn't ease this distress. While seeking proximity they also whined and continually nudged or pawed their owner for attention. Reunion did not bring security. It would appear that for dogs we certainly are an attachment figure, but the question remains: do they love us as we love them? Time to put them in the scanner (gulp!).

Dogs in the Scanner

I'd like to think he loves me. When you look him in the eyes you get that little dopamine hit and I like to think they get the same. Having cuddles makes him feel good. Being around him makes me feel good and being around me makes him feel good. **Matt**

All dog owners know the lure of the biscuit barrel to our four-legged, be-fluffed friends. My old dog Moose was apparently stone deaf when it came to doing anything he was told on a walk, or vacating the sofa when asked, but if we opened a biscuit tin several rooms away with the doors closed, he would be there instantly, puppy eyes and all. The argument is that dogs don't love us, they just love our food. In the UK we call this cupboard love, and debate has raged, mainly between confirmed dog owners and those of a more objective frame of mind, as to whether or not dogs really love us. For a long time we have had to rely on anecdotes and behavioural studies such as the one we encountered above, but it might be the case that we now have concrete neuroscientific evidence that dogs experience love too.

Oh, absolutely my dogs love me. Harley will nuzzle me, head-butt me, crawl under the blankets if we let her in the bed. For lack of a better term, she spoons with me. During the day the dogs are non-stop. Coming into the bedroom is a real treat, my wife is not keen on it. They will come in and they will do whatever they need to do stay in the bed. They cause no trouble. They lay down and army-crawl up next to us. **Russ**

The environment of the functional magnetic resonance imaging (fMRI) scanner is not conducive to relaxation. Wheeled into a narrow tube and subjected to a restrictive head restraint, patients or experimental participants are then assaulted with a cacophony of noise and banging as the three powerful magnets which are the basis of the fMRI scanner do their stuff. It is certainly the case that for some people this sensory assault is too much and sessions are abruptly aborted as panic sets in. Imagine, then, how impressed I was to read of the professor at Emory University in the United States who has trained dogs to lie still and unrestrained in his fMRI scanner, allowing him to draw conclusions about the brain of the domestic dog, and in particular its attunement to humans.

In 2016 Professor Greg Berns published his findings from his fMRI study on fifteen pet dogs, comparing their brain activity when offered a food reward as opposed to a social reward; in this case, human verbal praise. The dogs were veterans of the scanner and before the scanning experiment began, each dog went through a programme of training to teach them to associate the receipt of a particular reward with the appearance of a particular object. So, a toy car indicated the arrival of a food reward, in this case a piece of hotdog sausage, a toy horse the promise of some human verbal praise and, as a control, a hairbrush indicated no reward. After 120 rounds of out-of-scanner training, each dog was placed in the scanner and the three objects were presented at random at the end of a stick with the human experimenter out of sight. The use of objects meant that, in particular, the neural response to the food reward was not compromised by any possible social interaction with a human. If the car was presented – indicating a food reward – the dog received a food treat at the end of a stick. If the horse was presented – indicating a social reward – the dog's owner would appear at the end of the scanner and praise them; and if the hairbrush made an appearance, nothing happened.

What Greg's results showed was that in both the case of the food and social reward there was activation in the ventral striatum – home to the nucleus accumbens and caudate nucleus, which underpin the unconscious elements of human love and, as we know, are crammed with oxytocin and dopamine receptors. But, in all cases apart from two, the activation linked to social reward was equal to *if not greater* than that linked to a food reward. These dogs loved their owners *more* than food. In addition, as Greg and his team point out, the food reward on offer was one of the dogs' favourite treats – who doesn't love a hotdog? But the social reward was quite low-key in comparison; three seconds of verbal praise. Imagine what the level of activity would be like if dogs were petted, or petted *and* verbally praised together? Or what about a good session of tummy rubbing while snuggling on the sofa? It might be that when we look at the brains of dogs who are involved in the full joy of dog/human interaction we would see even more clearly that dogs love us not because we feed them but because they are as capable of love as we are.

We have this routine at night where she gets up on the bed with us and she takes turns kissing our foreheads, which I know grosses some people out, dog kisses on the forehead and ear. It's grooming and settling in for bed. She follows me around and it is not always because it is lunchtime! **Catriona**

So it is clear that dogs are attached to their owners and that their love for us stretches beyond our access to the treat cupboard. But as we know, attachment can be unreciprocated, so what about those of us who profess to love – to be attached to – our dogs? Perhaps one of the most powerful representations of the love a human can have for a dog is seen in the bond between the homeless and their pets. Despite struggling to have enough food for themselves and the severe discomfort and risk of living on the streets, many homeless people choose to have a dog, or other pet, despite it being another mouth to feed and an active barrier to finding a bed in a shelter, being able to take on a job or gain an education. But the relationship between pet and homeless owner can be, like many love relationships, of considerable benefit, which maybe goes some way to cancelling out some of the problems of pet ownership in this extreme environment.

Studies which have explored the relationship between homeless people and their pets have found that they play a protective role against depression, are often cited to be their owner's only source of companionship and love, act as social facilitators, encouraging the interactions with the public which lead to emotional and practical support, enable the establishment of a routine and, by caring for another, raise self-esteem and self-efficacy. Thus pet ownership has a positive impact upon mental health, and the evidence which suggests that it also reduces drug use – as peer influence is replaced by closeness to a pet – implies that these relationships have all the health benefits of those with fellow humans. In their 2016 study of the impact of pet ownership amongst homeless youth on rates of depression, Michelle Lem and her colleagues found that owning a pet reduced the risk of poor mental health by a third and pet ownership was a protective factor against substance abuse. For their eighty-nine pet-owning participants, many of whom left home before their sixteenth birthday, their pets *were* their family. The mantra on the street was one of 'pet before self', suggesting an almost altruistic, parental form of love underpinned by a powerful attachment between human and animal. And such a wide range of animals. Michelle's study listed the following 'participants':

fifty-two dogs, sixty cats, two rats, three rabbits, two bearded dragons, one chinchilla and a fish!

My dog is my kid. She is not our biological DNA obviously, but she is our developmental DNA. Her personality really fits. I struggle with some depression and anxiety and she gives me reasons to wake up in the morning. The same sense of mutual support. She is my buddy, she hangs out with me in the office, she is really good at keeping tabs on us if we are getting over-stressed about something. There is a lot more going on there than our culture gives credit for. I love my dog. **Catriona**

And as with others forms of love, this powerful inter-species bond is scaffolded by physiological and neurochemical mechanisms which will sound familiar to us all. A study of ten male Labrador dogs and their female owners carried out by Swedish scientist Linda Handlin found that blood oxytocin levels rose from baseline for both dogs and owners, but not for control humans, during an interaction involving stroking and talking to the dog for three minutes. Likewise, both dog and owners experienced a decrease in heart rate which endured for over fifty minutes after the end of the interaction, indicating a reduction in stress. However, when it came to the stress hormone cortisol, this was lower in humans as compared to baseline but not for dogs, perhaps because of the alien testing environment. While humans knew why they were in the room, the dogs didn't, and as all dog owners know, dogs find the unfamiliar the prompt for a considerable amount of panting and whining. These results mirrored those of a previous experiment, carried out by J. Odendaal and R. A. Meintjes that confirmed that interaction between a dog and their owner released a positive flood of neurochemistry including oxytocin, dopamine and, interestingly for the long-term nature of the bond between dog and owner, beta-endorphin. For dogs and their owners, the possibility exists that they share a complex and long-lasting love akin to that seen between ourselves.

I absolutely love my dog. I guess it's a feeling of happiness that applies to all my friends. We hang out and I am happy. **Sam**

It is difficult to say, although I think it is probably the case, that not all attachment is equal and some have a more profound impact on us, and are more powerful emotionally, than others, although I am sure there are some pet owners who would argue that the love they have for their pet, and the attachment security the relationship brings, is every bit as powerful as the love between parent and child. Indeed, a recent study found that overlap in neural activation did occur when women viewed images of their dog and their child – the fingerprint of love was there – although key differences indicated that it would still be the child you rescued first in a fire. But we might all agree that the attachment we are going to explore next is definitely not of the same league as those we have encountered thus far. Or maybe that's just a middle-aged mother speaking . . .

'Get Off Your Phone!'

Here are some statistics. 68 per cent of mobile phone users check their phone within fifteen minutes of waking; 10 per cent as soon as they wake up. 79 per cent of mobile phone users keep their mobile with them for all but two hours of their waking time. Two-thirds of users report distress on being separated from their mobile. The figure is significantly higher in young adults. 54 per cent couldn't be without their phone for more than two days. Americans check their phone on average every ten minutes. 13 per cent of millennials spend over twelve hours a day on their phone.

For many people, it sounds like the most significant relationship in their lives is with their phone.

Chris Fullwood is a scientist with a very modern research interest; the relationships people build with their smartphones. This makes his job title the rather cool-sounding cyberpsychologist. Chris is interested in how people engage with new technology and the impact it can have upon the social development of young people. And one area of his work which intrigues me is that relating to smartphone attachment. Just take another look at the statistics I presented above. The overwhelming need for proximity, the distress on separation, the desire to be in contact as soon as we wake and for as long as possible. Sounds like attachment behaviour, doesn't it? And if you hear the voices of Chris's participants it certainly sounds like love, with all its passions and exasperations:

I think whatever your relationship with them, love them or hate them, they are yours, aren't they? It's part of you, so if you can't find it, it is like there's part of you missing. **Joe**

This isn't a parent talking about their kid, it's a twenty-five-year-old member of a focus group explaining how he feels about his phone. Chris's work has found that for many people their phone is like another member of their social network. Where once they might have asked mum for a cake recipe, they now ask Google. Siri is the new best friend that finds restaurants and checks your spelling. Some have even mentioned that when a notification pings it is like the phone is crying out for attention. But is it the phone itself which is the focus of our attachment, or what it represents to us? The fact that studies have shown that any phone can assuage the distress of phone loss in the short term would suggest that it is not the physical object that elicits our attachment. Rather it is the fact that it represents the key method we now use to remain in contact with our loved ones, our real attachments, be they friends or family. It holds pictures of happy times, epic WhatsApp conversations and the possibility of real-time communication. For those who may be physically separated, it is a secure base to return to as it gives confidence that contact is only a call away. That is attachment in a nutshell.

I'm bad if I forget it. I panic then, because then I've lost contact, I've got no way of contacting anybody . . . I will go back home and fetch it, so I am sometimes late for stuff because I've had to go back for my phone. **Sara**

I am writing this chapter during the coronavirus outbreak of 2020, and having been a vociferous exponent of the supremacy of face-to-face interaction in all things love-related, I now find myself looking to my smartphone to maintain contact with those who I love but cannot be with at this time. While I have always argued that social media and modern forms of communication are just tools – we must control them rather than them us – it is amazing how important these tools become to us – we might even say they are part of our survival mechanism – during these times. So my mobile phone, while not an object of love in its own right, is precious to me right now because it is my lifeline to my parents and friends, without whom life would be pretty much unbearable. Maybe you feel the same way too.

If we were to find love with our mobile phone, then this would be an example of interspecies love. It is very unlikely that this would ever occur and if it did, the limitations of smartphone technology would make sure it went unreciprocated. But we are approaching another frontier in innovation which, some argue, presents the possibility that one day we might find friendship love, even romantic love, with a lump of metal and wiring. Artificial intelligence.

Enter the Robots

We don't have enough humans to care for us so it is not about what we want but what we need, but there are many, many ethical issues raised in this respect. The better the robot might be, the less human contact will be needed, but where is the limit? So there is a perfect robot and then I don't have to visit grandma at Christmas? The robot could look like me and have my voice! And I can go and do something else. That is the next level of the uncanny valley feeling. **Dr Marta Gácsi**, social-robot and canine scientist

In the west, a perfect storm of longer life spans, falling birth rates, austerity and geographical separation from extended family has led to a care crisis. There are simply not enough people, nor enough money, to make sure all of our most vulnerable members of society are adequately cared for. As AI has loomed into view and the prospect of a world shared with humanoid robots becomes a realistic proposition rather than the stuff of sci-fi, some have suggested that this new technology can provide the answer. A cohort of robotic carers. But caring is a behaviour backed up by a complex cognitive and biophysiological architecture. For to care isn't simply about cooking someone's tea, dispensing their pills or loading the washing machine. Caring is about walking through the door and realising that someone is having a bad day. It is about using your life experience combined with your empathy and theory of mind to know what to do next for the best, be it a hug, a stiff drink or a good joke to lighten the mood. It is all about the meeting of human minds, about biobehavioural synchrony. To build a caring robot would be an astonishing feat of engineering. But if it were possible, then as empathy and caring are the basis of love, is it possible that we could become friends with a robot? Could we love a robot and it love us back?

What is it about someone that makes us love them and them us? When it comes to human love, this is probably a very difficult question to answer. There will be elements of behaviour, of cognition, of psychology, of culture and religion. Some of these we will all share, but many will be unique to our particular circumstances, likes and dislikes. But to program a robot so that we can form, at the very least, a companionable relationship with it, we need to understand how these elements come together and result in love. So in answering this question it may be best to start somewhere simpler. Time to look at dogs again.

We argue dogs are a special species because dogs are really successful at communicating with humans even though they don't speak, even though they don't have a human-like face and facial gestures. It is tricky because in films and fiction you see highly developed robots . . . you could believe we could have super-robots. That is not true. There are robots who can sing, who really look like humans, there are robots that can walk, there are robots who can jump, but there is no robot that could operate in an everyday situation like a very, very dumb dog. **Dr Marta Gácsi**

Despite being less cognitively complex than us, dogs are highly competent social beings who exhibit both emotions and have distinct personalities and, as we know from above, we build powerful and beneficial reciprocated attachments to them that are full of love. As such, many who are in the vanguard of developing 'social robots' and acknowledge that developing a robot with the nuances of human social behaviour would be, if not impossible, technically prohibitively expensive at present, believe that dogs provide the perfect study subject as they try to break down what it is that makes us love our dogs. The first problem these people are up against is our natural suspicion – indeed, for some, extreme discomfort – of robots. This is the uncanny valley, which my interviewee Dr Marta Gácsi mentioned above, and is caused by beings who are very close, but not identical, to ourselves. It's why zombies are such a terrifying idea and good fodder for horror films. In her 2018 study entitled 'Should we love robots?' ethologist – that's an expert of behaviour – Veronika Konok explored the general public's attitudes to social robots: robots who would live in their homes. 176 men and women, 85 of whom had a dog, were asked a range of questions including their attitudes to domestic and companion robots, whether a companion robot could ease loneliness, whether robots might be a danger, whether and why they liked dogs and, the absolute clincher, is it possible to love a companion robot as much as a dog? The results were clear.

At present, 64 per cent of people would buy a domestic robot but only 13 per cent would go for a companion robot. The vast majority, 70 per cent, thought that it was impossible to love a robot as much as a dog, although 40 per cent felt they might be able to ease loneliness, a key aim when caring for the elderly. But when it came to what a dog does to make you love them, seventeen different attributes were listed, including being faithful, friendly, loving, cute, companionable and playful. Overall, dog owners liked their dog because they were attached/devoted to them, they had unique personalities or individuality and they loved them unconditionally. Interestingly, when it came to their negative traits, barking and stubbornness being two big ones, when owners were offered an identical dog but without these traits 92 per cent said they would not trade their dog in for a perfect version. 85 per cent of owners believed a companion robot could never be as good a companion as a dog because they have no emotions, no personality or individuality and, most nebulous of all, no 'soul'. In her conclusion, Veronika lays down the challenge of the companion robot. To pass the robot Turing Test – where an AI must pass as a human, i.e. be undetectable as a robot – they must be able to pass the attachment test, a very steep mountain to climb. But my interviewee, Dr Marta Gácsi from the Hungarian Academy of Sciences, argues that our quest to make our future helpers look human is where we are going wrong. We should not be trying to make a robot that looks like us or even like our pet dog – this, after all, increases the likelihood of the uncanny-valley effect – but let form follow function and if, to complete its tasks, it needs six legs and five hands, then give it six legs and five hands. Instead, we should be focusing on how to make a human attach to their robot; an inter-species attachment akin to that between dog and owner.

So it would appear the designers of our future robot companions have an uphill struggle to convince us that we need love with a robot. And this is made all the harder by our innate stubbornness in identifying a robot as anything other than an unrelatable alien. In attempting to understand how the human brain might perceive humanoid robots, a global team headed by neuroscientist Thierry Chaminade developed a robot with a human face and a range of complex facial and upper-body movements. WE-4RII – I think a rename would be on the cards if it ever moved in – could move its eyebrows, eyes, eyelids, lips, mouth, neck, shoulders and upper torso to perform four key emotional behaviours; anger, joy, disgust and speech. Before commencing their study, Thierry and his team made sure that human subjects could correctly perceive these four states by observing the robot's movements. They then placed thirteen participants in their fMRI scanner and showed them videos of a human and WE-4RII performing the range of emotions. Participants were asked to say which emotion was being displayed for each clip. The results showed that in some areas the brain showed more activity when the robot was viewed than the human. These differences were particularly strong in the visual areas of the brain as the brain had to undertake increased visual processing to interpret the face of the robot. It was being read as an 'alien' being. But in other areas the activity when viewing the robot was not much different from baseline and most critically both the mirror neurones which fire when we see the movement of a fellow human and the social areas of the brain in the orbitofrontal cortex were inactive. The brain quite clearly did not perceive the robot as human and showed no motivation to want to make social contact with it, a major problem if we are to rely on these robots to care for our elderly or consider them a potential romantic partner.

The most important aspect of robot development is attachment. They need to show 'you are my friend, my mother, a relative'. We need robots who can fulfil this social interaction, the affectional aspects of interactions and give concrete physical help at the same time. In one of my studies we studied a group of disabled persons who have assistant dogs to try and understand why they accept the assistant dog more than the assistant robot, even though an assistant dog would never be able to bring them a glass of water because of the physical constraints. **Marta Gácsi**

So where does this leave robot love? At most, it would appear that we may experience an inter-species love, although at present we cannot make this love as powerful as that we share with our dogs. And there would be a lot of cultural hurdles to cross; our deep-seated suspicion is not going to go away overnight. We are happy to have a robot hoovering the floor or mowing the lawn, but as for the hallmarks of human friendship – holding our hand and sharing our most private and intimate moments – not so much. But what concerns me is that those for whom the human-oid robots will be first developed, those requiring care, will not get a choice. Because if a government can solve the care crisis by replacing our overworked human carers with robots, they might just do it. Yes, there are initial development costs, but once up and running the costs of running a robot – a bit of maintenance, the odd upgrade – will pale into insignificance next to the costs of paying for human carers. Robots don't get ill, they don't take holidays, they don't leave to go to other jobs and they can work 24/7. And this may leave our most vulnerable in society with a level of practical care but none of the advantages that come with real human care, those which have a direct impact on our health, happiness and quality of life. There would be no meeting of human minds; no biobehavioural synchrony. The ideal scenario would be for a robot carer to do all the dull practical stuff and free up the human carer to sit, hold hands and chat. To be a much-needed friend. But I am not optimistic that in a world of austerity this is what the outcome would be. So as with so much of 'future love', there are ethical questions we need to consider. We need to ask ourselves, would we want our grandmother, our disabled sibling or our new baby to be cared for by a robot? And if not, how do we make sure AI is not allowed to enter the most intimate areas of our domestic lives?

I am not scaremongering. As I write this in September 2020, a study by the University of Bedfordshire has been released which claims that humanoid robots introduced into care homes in the UK and Japan have led to increases in good mental health in their inhabitants and decreased loneliness.[2] The study has been met with overwhelmingly positive coverage in the media – tempered slightly by a thunderous article in the *Guardian* newspaper from me – although the general public has been less keen. And if we were to believe the outcomes of Marta Gácsi's research, robots who can care effectively and without inflicting harm are a long way down the line. It is incredibly concerning to me that they are being introduced into care homes when they bear none of the hallmarks of behaviour that would lead to the attachment which is fundamental to healthy social interaction, be it with our fellow humans or our dog. I am not saying that AI has no place in social care – that would be like trying to do a King Canute and hold back the tide – but we have to be very aware of what we will and will not accept when it comes to the care of the most vulnerable people in our society, and we can only do this from a position of complete knowledge. A debate is needed and, as roboticists surge ahead with their plans, it would be best if this happened sooner rather than later.

---

2   Just a note here that this study ran for just two weeks, meaning that this positive finding may simply be a result of novelty rather than any true indication that interacting with a robot instilled the same benefits to mental health as that we receive from interacting with a fellow human or, indeed, a dog.

This chapter has been about looking at the love we underestimate, that we feel for our friends. We have learned that friendship love is an increasingly important aspect of our lives as our world evolves and fewer of us find romantic love or have children. We know that friends are our source of solace, unjudgemental support and fun, all of which is underpinned by a meeting of similar minds. We have met the concept of chosen family and the idea that our fictive kin are as much about survival as our biological. We have delighted (I hope) in the love we have for our dogs and learned that they do genuinely love us back, it's not just the biscuits. And we have glimpsed a future where the rise of AI might lead to some of our closest friends being made of metal and wires. But beyond all of this, I hope that this chapter, and the one which precedes it, has given you a glimpse of the huge spectrum of opportunities for love we have in our lives. Sometimes we just need to lift our heads a little higher to see it.

CHAPTER FIVE
PERSONAL

● ● ● ●

What is love? Here are some thoughts from an Oxford audience:

'To love someone is to value them. Sometimes more than yourself.'
'. . . a shared burden and freedom to brace against the tempests of the world and share in its joy.'
'My three cats.'
'Love is warm, comforting, happy, exciting, carefree, unconditional, peaceful, communicative, creative.'
'Love is a) a survival mechanism b) an invention by Clinton's cards.'
'A figment of the imagination that bamboozles one's brain!'
'Comfortable.'
'No clue – but what else would be the purpose of life but to love whoever is around to be loved?'
'Love is the partner to fear.'
'Love is the result of multiple chemical reactions to help us attract a mate for the primal instinct of procreation.'
'My kids.'
'A package of warmth, interest, compassion and a tasty butt.'

I could go on . . .

The genesis of this book came about in 2017 when I took part in a public engagement event at the University of Oxford, where I was based. I was asked to lead a debate with members of the public based upon the question 'What is Love?' Before the evening commenced, I asked each member of my audience – around 300 in total – to write down their thoughts about the answer to this question, anonymously, on a piece of paper. Above are just some of those responses. It brought home to me the immense subjectivity of love. How some people associate love only with romantic feelings, while others extend it to cover kids, pets and objects. How some describe it in terms of a powerful emotion, while others believe it to be just a construct dreamt up by those who want to sell us stuff. How some feel the question merits an essay while others feel capable of answering the question with one word. And how some people simply have no clue what love is but know that it is essential to the experience of life. But one thing was for sure: on that September evening in Oxford, everyone had an opinion.

One aspect of love that we can be certain of is its subjectivity. I will never know if how I feel when I am in love is the same as how you feel. In all likelihood, such is the complexity of love, we do not. And while we may not be able to accurately quantify this difference, what we do know is that how we feel and act when we are in love is down to a mix of factors which are exclusive to us as individuals. Our genetics, our psychology, our biology, our culture, our life experience, plus, of course, the mysterious ingredient X which is unique to just you or just me.

Having studied thousands of relationships, you'd think I would be pretty good at spotting a good one by now, to be able to predict the future outcome of a nascent love. But no. The frustration and joy of studying humans is that they always throw you a curveball, making a choice or acting in a way that is entirely unpredictable

– the committed environmentalist who shirks the large body of research on relationships that dictates similar values are a must and befriends the hunting enthusiast, or the adrenaline junky who is unbothered by our pronouncements that shared activities are a good thing and falls in love with the homebody. But as my colleagues and I come to understand the elements of love, we are also starting to understand where some of the individual differences in sensation and experience come from.

I don't think love can be explained in a simple sentence. I don't think you can say what I feel for someone is the same as what someone else feels. I don't think you can say it is the same for societies as a whole, across countries, across cultures ... For me, it is like the feeling of being home. **Charlotte**

In this chapter we are going to explore how your genes, your biological sex and gender, your age, your ethnicity and your sexuality may play a role in how you experience love. We are going to encounter the genes which give some an armour-plated shield, allowing them to defend themselves against the effects a harsh upbringing could inflict upon their abilities to give and receive love, and the genes which underpin the sensitive parenting which allows our children to thrive. We will explore the sensitive topic of whether our ethnicity influences the nature of the love genes we carry and whether we can see this in the characteristics that we ascribe to different populations. We will highlight the hot topic of gender and how our gendered ideas about how men and women should behave when in love infiltrate our minds at an early age, and we will enter the ongoing debate about whether men and women experience love differently. We will hear the voices of my interviewees, who have all provided me with their thoughts on what love is. This is a chapter all about the individual experience of love. But first, a quick look at how genes work.

Genetics 101

Love is a combination of a feeling of trust and security, comfort. A feeling that not only are you appreciated but that everything that love means to you is being reciprocated by the target of your affections. A true love is when what you give to your target of affection is 100 per cent reciprocated by the other person. Without doubt. **Russ**

We have already encountered one of the factors that can influence our individual experience of love in Chapter 3: your attachment style. And we know that beyond a minimal genetic input, this is largely shaped by your environment, particularly when you are developing. But at the other end of the nature/nurture spectrum are your genes, and the research that we carry out at Oxford, combined with that of our colleagues around the world, is showing us that the way you love does have a considerable genetic component. But before we dive into a genetics 101, here's the disclaimer. Genes are not deterministic. We often describe a gene as being 'for' something, but this is a gross inaccuracy. Genes not only interact to a greater or lesser extent with the environment but also with each other, so the most we can say if you carry a certain gene is that it increases the chance of the particular attribute or behaviour it is associated with expressing itself in you – the phenotype. Some genes have a greater influence than others – the BRCA1 gene increases the chance of developing breast cancer by between 50 per cent and 85 per cent – but very few have absolute control on how you will turn out. The more complex the attribute or behaviour, the greater the number of genes which influence it and the less powerful any one single gene is in determining outcome. The genetics of love definitely reside in this category. But one of the genes that might have a bigger role to play than some others in the game of love is the oxytocin receptor (OXTR) gene.

The OXTR gene is a highly variable beast. In technical terms, it is polymorphic. This means that because many sections of the gene can differ between individuals, it comes in many forms, and because of this it seems to have a disproportionate influence, compared to the other genes which influence the neurochemistry of love, on how you feel and behave when in love. The most studied of these differences are the SNPs – the single nucleotide polymorphisms – and OXTR has *a lot* of these; twenty-eight at the last count. The nucleotides (the 'N' in SNP) are an important element of DNA and they come in four different types; adenine (A), guanine (G), cytosine (C) and thymine (T). At the most basic level, SNPs occur when these bases vary between individuals and on the OXTR gene there are at least twenty-eight different places along the gene where this happens. This results in people carrying different versions of the same gene and when something has at least twenty-eight SNPs (we haven't finished looking yet) that's a *lot* of different versions. We are going to meet some of these SNPs in a little while.

But how does a gene influence what is actually happening in us? When it comes to a gene associated with a brain receptor, like OXTR, it can be in a few ways. First of all, it may influence the actual number of receptors you have in a particular area of the brain – the density. The argument would be the greater the number of receptors, the more powerful the impact of the neurochemical on you. Or it might influence where they are. So, again, if you have more oxytocin receptors in your limbic area – remember, this is where attraction starts – it might make you more open to starting a relationship. Or the gene may influence the affinity of your receptor for its partnered neurochemical (for OXTR the partnered neurochemical is oxytocin). If you imagine the receptor and neurochemical as a lock and key, the neurochemical – the key – needs to latch on to the receptor – the lock – to enable its message to continue down the neural pathway. This fit between receptor and neurochemical can be on a spectrum from not that great to really tight. The better the fit, the better and more efficiently the message is communicated around the brain. And the more quickly, and maybe more intensely, the neurochemical will have its effect upon you. But how good this fit or affinity is is partly dependent on the gene which encodes for the receptor, so for some people the affinity between the OXTR and oxytocin, for example, will be greater than for others. For some the lock will ease open, while others might need to jiggle the key a bit. Beyond the genes for the receptors, your genes can also influence your baseline levels of the neurochemicals. We all exist at different baseline levels of the neurochemicals which underpin love, and it is in part this difference that impacts why we all experience love differently. For example, the lower your level of baseline oxytocin, the less likely you are to be open to starting new relationships. Lastly, your genes can influence how efficiently neurochemicals are transported around the body – not all neurochemicals are made where they are needed. Oxytocin is manufactured in the limbic area's hypothalamus but released by the same area's pituitary gland. Our genes can influence how good our brain is at shunting oxytocin from one area to the other, and in oxytocin's case sluggish transport might mean a delay in its vital confidence-building benefits.

Love is being willing to sacrifice or endure hardship for the benefit of another. Willing to do anything for them. Without question. And there is a sliding scale of that. There are certain things I would do for my friends but there is practically nothing I wouldn't do for my son. Love is selflessness. **Matt**

Because of its variability, OXTR has been posited to have a wide-ranging impact on our love and prosocial behaviours, encompassing the full range of relationships from lovers to parents, friends to communities. It has been suggested that it influences how big our social networks are, how likely we are to suffer from social anxiety, how sensitive we are at parenting, how strong our empathising skills are and how easy we find it to trust. One of my favourite early discoveries was SNP rs7632287 – a bit of a mouthful so I nicknamed it the 'nesting' SNP. Remember that an SNP occurs when people carry different nucleotides – guanine, adenine, cytosine or thymine – at a particular point along a gene. They are variables on a gene that we all carry. When it comes to our nesting SNP, those who carried a guanine (G) base in the critical position – making them GG (remember chromosomes come in pairs, so genes do too) – were much more likely to be in a pair bond and derive greatest satisfaction from it compared to those who carried an adenine base in the same position (either AG or AA), who were more likely, if in a pair bond at all, to be in relationship crisis and report low relationship satisfaction. But as time has gone on we have been able to broaden and replicate our studies and this has led to two SNPs appearing to have a more profound impact on our love than any of the other twenty-eight at present. These are the snappily named rs53576 and rs2254298. The body of evidence we now have for their influence enables us to draw strong conclusions about their impact on our individual experiences of love with our lovers, children, friends and community. And not only does the influence of your genes change with age, but your ethnicity may also impact how likely you are to carry a particular version of a gene.

The Genetics of Empathy

By now it should have become apparent that empathy sits at the basis of all our relationships and is critical for our interactions with others, critical for love. In the early days, those who studied the genetics underpinning our neurochemistry – pharmacogenetics – regularly concluded that the OXTR gene, and the rs53576 SNP in particular, had a key role in the extent to which an individual could display empathy. However, these studies were often dogged by small sample sizes and too many variables – different age groups, sexes, ethnicities – meaning that the results were often still murky at best. So in 2017 a team from China led by biologist Pingyuan Gong decided to carry out a two-stranded study with the aim of putting the debate to bed once and for all. In the first strand they recruited 1830 college students, the majority of whom were female and all of whom were young (having an average age of twenty years) and of a single ethnic origin, being Han Chinese. They asked the participants to complete an established measure of empathy – the Interpersonal Reactivity Index (IRI) – and took their blood so they could be genotyped for the version of the rs53576 SNP that they carried. What they found was that the identity of the base at the SNP site – in this case, A or G – could predict an individual's IRI score. The higher the number of G alleles you carried – remember, the options are GG, AG or AA – the more empathetic you were. A strong result based on a good-sized and well-controlled participant pool. But not happy with just their result, they then set about pooling these results with those of all the other studies – twelve in total – which had explored this question before. This is known as a meta-analysis and it means that we can test our hypothesis on a much larger pool of data, and control well for confounding variables, more than we ever could in a single study. Taking the resultant pool of 6631 participants, they confirmed that the allele of rs53576 an individual carries has a significant impact on empathetic skill. Those who were homozygous for G – i.e. GG – displayed the most level of skill – meaning there appeared to be a dose response. A dose response is when the dose (in this case, empathetic skill level) is related to how many units of something you have. So here it is how many G alleles you have. Two gives you a higher impact than one. Beyond this, because their meta-analysis pooled data from studies in Asia and Europe, they were able to show that while rs53576 had an equal impact on empathic ability regardless of your ethnicity, there was a difference in the likelihood of carrying a G allele, dependent upon where you were born. So European populations have a significantly higher frequency of the G allele than Asian populations, and the authors supported their conclusion by referencing the psychological literature that showed Europeans tended to employ empathy to maintain their relationships to a greater extent than those of Asian background. Sociology has shown us that the reason for this may be linked to the differing natures of the two societies. Asian populations tend to be more collectivist – focused on what is for the good of the group – than individualist – the hallmark of most European populations – indicating that the ability to tune in to the emotional state of a single other is of significantly more importance in the European population than the Asian. We'll return to the role of your ethnicity in your experience of love a little later in this chapter.

Armour-plated Genes

Love is a unique, intangible connection with another being. It's something that fills your heart and soul in a way no other connection can. **Gemma**

Often when I speak about genes I am aware of a sense of discomfort or alarm making its way through the audience that stems from a dislike of the idea that we might be, even a tiny bit, at the mercy of our biology. That even with our massive brains and apparent abilities to overrule our baser instincts there still might be aspects of who we are that are not open to self-determination. That we are, at the end of the day, just another animal. This is one way to look at it, but as there are two sides to every story, so there are two sides to this particular genetic tale. Yes, we might have a set of genes which influence our path in life in a less than healthy or satisfactory way – maybe some of those linked to addiction are prime candidates – but there are also genes which act as our armour plating, protecting us from the sometimes extremely damaging effects of our environment. Take the impact of childhood abuse or neglect. As we know from Chapter 3, how we are cared for or parented as small children can impact the very structure of our brains, particularly in the prefrontal cortex where our social cognition, and the cognitive aspects of love, find their home. For many years now, I have been left speechless by some of the stories I have heard about my participants' childhoods. Stories of extreme emotional, physical and sexual abuse and neglect. For some of these people, the statistics that link such an environment to increased chances of psychopathology, addiction and social dysfunction have become their reality, but there are always some people who seem to have dodged the bullet. Who have matured into healthy, socially competent and happy beings. When asked why, I have often responded with my untested and rather generalist belief that they must have a cast-iron set of genes. That there is something innate within them that makes them more resilient than most to the deprivations of their developmental environment. And very recently one element of this genetic armour has been identified as an SNP we have met before: rs53576.

Abuse comes in several guises; sexual, physical and emotional. While it is relatively clear what types of behaviour are categorised by the first two, it is the latter type of abuse which is the hardest to define, to recognise; but it is this form of abuse, it is argued, that has the greatest impact upon our social and emotional development. Children who have experienced this form of abuse tend to be less competent at forming and maintaining relationships as adults, are less able to navigate social interactions successfully – that's knowing the rules and displaying the skills of the social game – and many, rather than viewing their close relationships as a source of comfort and support, view them in a negative light. This group are more likely to display destructive externalising or internalising behaviours, to experience poor mental health and to have difficulty with normal emotional function which may impact their ability to form loving relationships. However, there are some for whom such problems are not an issue. In their 2019 paper published in the journal *Child Abuse and Neglect*, psychologists Ashley Ebbert, Frank Infurna, Suniya Luthar, Kathryn Lemery-Chalfant and William Corbin explored the question of why, for some, a childhood of emotional abuse was a one-way ticket to a lifetime of issues with close relationships and why others defied the statistics. Further, they wanted to understand whether the impact of early abuse had more of an effect on some types of future relationship – those with family, lover or friends – than others.

Ashley and her team recruited 614 participants from a long-term study – the AS U Live Project – which explores resilience in mid-life. The participants were asked to complete question-naires which asked them about instances of trauma during their childhood, including emotional abuse, and about the quality of their current relationships with their friends, family and romantic partner. Participants' genotypes for the rs53576 SNP were assessed from blood samples, with results showing that 279 individuals were GG (homozygous for G), 265 were AG (heterozygous) and 69 were AA (homozygous for A). In the first instance, the study's results showed that your genotype for rs53576 did not influence the likelihood that you experienced abuse as a child; instances of abuse were evenly distributed across the three allele groups. How-ever, if you were homozygous for G – that is, GG – and experienced emotional abuse as a child, you were significantly more likely to report satisfaction with and support from your adult relationships with family, friends and lovers. In contrast, those who carried an A allele, that is AG or AA, were more likely to report that these relationships were strained. These results aligned with previous studies which showed that individuals with the GG genotype tend to show better psychological adjustment, accurate processing of social information, increased trust and supportive relationships. What is it about the GG genotype that enables these individuals to have had a better outcome regardless of their difficult start in life? The team behind this study suggest that, as A carriers tend to look on social support more negatively than GG carriers, it may be that GG carriers experience a greater neurochemical reward from their social interactions than A carriers, which encourages them to work to build and then maintain these vital survival-based relationships.

Extraordinary Epigenesis

Love is a state of mind where logic falls apart. **Jim**

Ashley Ebbert's study focused on a group of people who were some way down their life road, and there is some suggestion that, regardless of your genotype, as we age the impact of our child-hoods on our social functioning and our abilities to experience healthy love diminish as we gain more life experience and come to recognise and work against those behaviours that may be less than helpful to our aim of a happy and healthy life. As a consequence, our age is one of the factors which impacts our individual experience of love. Even personally, I know that my experience of love has changed as I have moved from the crush-laden teenage years, to early explorations of first love at university, through issues with attachment at the start of my marriage to the intensity of parental love and the companionate love, with the odd flashes of passion, which characterise a marriage of twenty years, two children and eleven pets. And the statistics support the idea that the younger we are, the more likely we are to exhibit attachment anxiety, possibly linked to the instability of our relationships and social network as we transition from school to university to work, combined with a shift in attachment focus from our parents to our friends, and ultimately to our own long-term partner and children. But our genes, and the mechanisms which regulate their operation, and ultimately expression, may also have a role to play.

When I first encountered genes and inheritance at school, the mantra was that at the moment the sperm met the egg and the two sets of chromosomes came together to make the first single cell, our genes were fixed for life and these genes would be passed, untouched, to our children. Nothing we did during our lifetime – our environment – could alter them, meaning that, however hard and fast we lived, our sins would not be visited on our offspring, genetically at least. But in recent years this story has started to unravel, and one of the first hints that it might not be as rock solid as we first thought came from Swedish history.

In nineteenth-century northern Sweden, the population lived under a cycle of crop failures and crop abundance, which meant, during times of plenty, they would take the opportunity to indulge themselves in some serious overeating. Two generations later, it was noticed that the grandchildren of these people had a much higher risk of dying from diabetes-related causes and heart disease than the average population. It was as if they had been enjoying a very rich diet, even though they had done no such thing. Why was diabetes, usually a disease of the obese, so prevalent in this population? The answer lay decades in the past. The environment of plenty that their grandparents had periodically enjoyed was having its impact two generations later, on their grandchildren's health. The expression of the grandparents' genes had been altered by their dietary overindulgence and this had been inherited by their children and then passed on to their grandchildren. This is epigenesis.

Epigenesis does not describe an alteration in the genetic code itself – that's the DNA – rather, it is an alteration to the way the gene is operated. It is a bit like the DNA being the hardware and the epigenesis being the software altering the way the gene is expressed. So, for example, the chromatin, tasked with condensing

DNA so it fits within the cell nucleus, or the histone – the protein around which DNA wraps itself – or whether a gene is 'on' or 'off' (otherwise known as silenced) are altered rather than the genes themselves. The software is modified rather than the hardware. And it is becoming clear that epigenesis of the OXTR gene is one of the mechanisms which impacts how we love, and can be one of the mechanisms that causes the transmission of social skill, including our abilities to love and be loved, across the generations.

Ultimately the role of DNA and our genes is to code for and then manufacture the proteins which are the foundation of all life. DNA methylation describes the process by which a methyl group (CH3) is added to the DNA, which alters the function or expression of the gene. Methylation varies between individuals and one of the factors which impacts the degree of individual methylation is epigenesis. A recent meta-analysis of thirty published studies exploring the impact of methylation of the OXTR gene on psychological, neural and behavioural outcomes by a team of psychologists and psychiatrists, led by Eline Kraaijenvanger, allows us to be confident in our understanding that this process does have an impact on social functioning. In particular, methylation of one particular area of the OXTR gene – which limits the expression of the gene and by association our social abilities – has been associated with a number of outcomes which impact how we behave in, and experience, our close relationships, including social functioning, social-behavioural disorders and poor mental health. Indeed, brain scans of individuals who have high degrees of OXTR methylation within this region – CpG (cytosine-guanine) site –934, for those who like the specifics – show heightened activity in the prefrontal cortex, indicating that for these individuals engaging in social interaction is much more effortful than for those with less methylation. But as experience enables us to find more peace within our relationships as we age and leave the tumultuous early years behind us, so age may reduce the impact of DNA methylation upon us and lead the way to more secure attachments.

Love Across the Life Course

One of the consequences of working in a university with an apparently endless supply of student guinea pigs is that the vast majority of research is carried out on twenty-year-olds eager to gain course credit. This is of course wonderful, but it does mean that assumptions can be made about adult development which don't hold true as we age. In 2019 psychologist Natalie Ebner and her team from the University of Florida decided to buck this trend and explore the link between OXTR methylation, oxytocin levels and attachment across the life course. They recruited twenty-two young people, mean age of twenty-two, and thirty-four older people, mean age of seventy-one, all Caucasian with around a 50:50 sex split. They asked both groups to complete an adult attachment questionnaire, the Experiences in Close Relationships measure which we encountered in Chapter 3, and took blood to allow analysis of levels of methylation at site –934 – known as epigenotyping – and baseline oxytocin levels. What they found was that while both groups exhibited comparable ranges of OXTR methylation and baseline oxytocin levels, it was only among the younger cohort that there was any link between these and attachment style. So within this group, the lower the methylation and higher the oxytocin level, the less likely it was that the individual would exhibit attachment anxiety. In contrast, the degree of methylation seemed to have no impact on attachment style in the older cohort. This would suggest, the team concluded, that as we journey through our life course the environment of our ancestors, as reflected in the epigenetic phenomenon of DNA methylation, has less of an impact on how we feel and behave. Instead, the influence of our own environment comes to the fore. However, as epigenetic alteration of gene expression is heritable, this does not mean that any offspring of this older generation will not feel the impacts of OXTR methylation on their own love experiences, particularly when young, just as the grandchildren of the Swedish farmers felt the impact of their over-indulgent grandparents. It simply means that as we age and gain experience and confidence, the impact of our genes lessens to some extent.

The Genetics of Mum and Dad

Love is when what happens to them matters to you as if they were extensions of yourself which you care for even more than you care for yourself. **Sean**

In my other life, I have the privilege of following men and women as they become parents for the first time. I meet them first when they are excited parents-to-be, full of ideas about how they will parent and who their child will be, then again within the first weeks of their baby's life, when it is hard not to compare them to a rabbit caught in the headlights, and then onward through their baby's first years as they navigate the rollercoaster of early parenthood. For many, creating the perfect developmental environment for their child is of primary concern and they understand that, particularly in the early years before their child ventures beyond the home to the semi-independence of nursery life, the buck largely stops with them. However, as their child's genes will dictate to what extent the parents' carefully curated environment will influence their baby's social development – remember some genes act as an armour, reducing the impact of a negative environment – so how they parent, and as a consequence the environment they go on to build for their child, is in part influenced by their genes. It's time to meet a couple more OXTR single nucleotide polymorphisms (SNPs) – rs1042778 and rs2254298 – and their closely associated colleague rs3796863, which is found on the CD38 gene and is associated with the release of oxytocin from its site of manufacture in the hypothalamus. All three have an influence on how we parent, the strength of the loving bond we are capable of building with our child and, ultimately, the love our child will feel for their first best friend.

Those who work with families know that one of the key factors impacting whether or not a parent will struggle to bond with their own child is how they were loved and cared for as a child. In particular, neglect crosses generations and is a vicious cycle which is difficult to break. To understand why this is, we must begin by trying to understand the underlying mechanism of this phenomenon and to do this we can't just dip into a child's life for a moment and take a snapshot of their home life, we must follow them, and their parent, on their developmental journey. In 2013 Ruth Feldman and her team at Bar-Ilan University in Israel did just this as they sought the key to explain how a parent's behaviour, influenced in part by their genes, can shape a child's social world as they enter the pre-school setting. They recruited fifty cohabiting, heterosexual Caucasian couples and their one-month old infants. During this longitudinal study they carried out two infancy visits, when the babies were one and six months old, and one pre-school visit when the children were three years of age. On the first visit at one month, blood was taken from both parents to allow their baseline levels of oxytocin to be assessed and to enable genotyping to occur for the three SNPs of interest – two for OXTR and one for CD38. Parents were then asked to play with their child as they would usually and the sessions were videoed. At the six-month visits a parent–child play session for both parents was again taped. Fast forward a few years and families were re-visited when their children were three years of age. At this instance the protocol was more extensive. Salivary oxytocin – drawing blood from children for non-medical research is ethically a no-no, but the two measures are comparable – was collected at baseline and after a playful interaction for mum, dad and child. Following this, the child was invited to play with their best friend from pre-school – a structured session involving an animal farm – for ten minutes.

Throughout the study, the taped sessions were carefully coded for behaviour. During the infancy sessions four categories of parenting behaviour were assessed: gaze – including infant-directed and gaze aversion; affect – positive, negative and neutral; vocalisations – 'motherese', general and none; and touch – affectionate, functional, stimulatory and none. At three years of age the coding was more complex, including as it did the observation of parental and child behaviour, but was based upon instances of social reciprocity such as the sensitive give and take of verbal and non-verbal cues, adapting to each other's needs and engaging in shared activity.

What Feldman's team showed was that parents who carried the CC allele – labelled a risk variant – of CD38 rs3796863 exhibited not only lower levels of oxytocin but lower levels of sensitive parental care than those who carried the A allele. This meant that these parents showed more negative or neutral affect, maintained less infant gaze and indeed may have actively avoided it, exhibited less sensitive touch and carried out less positive, infant-directed vocalisation. For mothers, this relationship also held for the risk variants of OXTRrs1042778 (TT) and OXTRrs2254298 (GG). It was clear that the parents' genes were impacting how capable they were at parenting sensitively and, in turn, how secure the attachment – remember, attachment is the objective measure of love – between parent and child. When it came to explore the results for the assessments at pre-school age, a child's baseline and post-interaction oxytocin levels were correlated with those of both their father and mother – this is a consequence of biobehavioural synchrony – but when considering the reciprocity of the interactions with their best friend, it was only mum's parenting behaviour during infancy that impacted the positivity of these interactions. So while both mum and dad's oxytocin levels, which are influenced by the version of CD38 they carry, shaped those of their child as they entered the world beyond the family, it is mum's love that shapes our first friendship loves.

But beyond the role for these genes in our individual stories of parenting, Feldman's conclusions show that, alongside epigenesis, one of the mechanisms by which our abilities to show and experience love across generations is the impact that our parents' genetic make-up has on our environment of development. We know from Chapter 3 that a parent's brain structure actively shapes their child's social competencies and with this study Feldman adds to that literature; parents' genes, and particularly those of mum, actively influence the special relationship they have with their very first close friend. Further, her study highlights that there are periods of time in a child's life which are of particular importance in forming a child's prosocial competencies. The first six months of a child's life seem to be critical for developing the high levels of oxytocin which will enable them to form functioning relationships throughout their life, but the love they experience with their first best friend may also be an arena in which interventions can be made to turn the tide of cross-generational neglect.

Love and Ethnicity

Love is the general connection we have with other people. It has many different shades and many different forms and we have tried to categorise it by trying to categorise romantic love, or familial love, or platonic, and it doesn't quite fit. Sometimes it appears slowly and other times you meet someone and it is 'Wow, they are incredible. I adore them.' It allows contradictions to exist in the same space. **Bill**

So far, we know that genes come into our individual experience of love on several fronts, via the conventional routes of genetic inheritance and via the more recently recognised mechanisms of epigenesis. That our genes influence all areas of our love experience – romantic, familial, parental and platonic – and can act as a suit of armour allowing us to sail through negative developmental environments with our ability to love unharmed. And that both our age and our ethnicity have an influence on whether or not a gene is expressed or even the likelihood we will carry a particular version. We first encountered the issue of ethnicity earlier in this chapter when I looked at Chinese biologist Pingyuan Gong's meta-analysis on the genetics of empathy among European and Asian populations. She concluded that the G allele of rs53576, an OXTR SNP, was at significantly higher frequency in the European population than the Asian population, which reflected what we knew about their social psychology. As I mentioned above, this finding reflects conclusions from sociological research that Asian populations tend to be more collectivist than individualist, meaning that empathy for individuals is a skill which is employed to a greater extent in the latter population, because it is more relevant, than in the first population.

Recently, a Russian team have extended this work to a broader range of populations from Asia, Africa and Europe, with the aim of mapping global frequencies of rs53576 and rs2254298, which, as we know, have been linked to empathising and parenting behaviours. The team, led by geneticist Polina Butovskaya, collected saliva samples for genotyping from four different populations; Siberian Ob-Ugric (353 people), Tanzanian Hadza (135 people), Tanzanian Datoga (196 people) and 659 Catalonians from Barcelona. They then compared their results to a subset of the 1000 Genome database which contained the results for fifteen further populations spanning Africa, South America, Asia and northern Europe, enabling them to place their findings within a broader global context. What they found was that, with respect to rs2254298, which underpins both the ability to sensitively parent and empathise more broadly, the frequency of the A risk allele was at its lowest in the European population. Asian populations showed the highest frequency, with African populations sitting at the mid-point between the two. Where rs53576 was concerned, Africans and Europeans had significantly lower frequencies of the A risk allele than Asian populations. This would imply, at the most broad brush of levels, that European and African populations had more competent empathisers within their populations than Asian populations, while where sensitive parenting is concerned the range ran from a high genetic tendency in Europeans, through Africans to a low tendency among Asians.

As I write this, I am aware that this is pretty explosive stuff. Are we really suggesting that your ethnicity impacts your capacity for love or how you behave when with those you love? In the first instance, this idea makes me incredibly uncomfortable and is the stuff that those who wish to drive a wedge between ethnic groups leap on with glee as ammunition for their hateful arguments. But if we dig a bit deeper, the story is not so clear cut. In the first instance, if the A allele is a risk allele – implying it bears a disadvantage to survival – why does it exist at such high frequencies in some populations? There may be two main reasons. The first is that cultural practices may have skewed the natural order as a result of selective breeding or investment. Secondly, it might be that the A allele confers an as yet unidentified advantage on its carriers which balances out the impact it may have upon empathising or sensitive parenting. The research which suggests that carriers of the 'beneficial' G allele are more likely to suffer depressive disorders – which in themselves impact one's ability to experience and show love – emphasises that the dichotomy between risky and advantageous alleles is not as clear cut as it would first appear when it comes to love. And do not forget that genes are not deterministic, complex interactions with other genes and the environment mean that what is written in our genotype isn't necessarily expressed in the phenotype.

This lack of clarity should not surprise us. I hope it is becoming clear that love is a many, many-faceted phenomenon and no one factor – our ability to empathise, our mental health, our upbringing, our genes, our age, our sex, our oxytocin levels or even our waist-to-hip ratio – has the steer when it comes to our lived experience of love. There are many genes which underpin empathy and there are many ways to express parental love. While having the genes which promote sensitive parenting might be advantageous to our children's development in a Western context, it might be a less efficacious set of skills in an environment, say, where day-to-day survival is at risk. In this instance, being able to pick your kid up and hurl them over your shoulder while legging it from a threat is a more powerful representation of your love than holding their gaze and gently explaining what is going on while stroking them, by which time the lion has arrived to eat you. When it comes to love, the context is all-important, and where we might find a deficit in skill or feeling here, we will find an abundance there, and it is that balance – which is individual to us all – which will determine our experience of and ability to love.

Gendered Love

Love is both a mystery and a kind of madness. I think it's where there is a very intense preoccupation with somebody or something. I use the word love to talk about an intensity of feeling, of being drawn to, unable to let go of, an inner preoccupation with . . . Love is a kind of psychosis. **Benjamin**

While the broader results of Polina's study made for thoughtful reading, one surprising finding did pop out: there was no difference in allele frequency between men and women within their study populations. We know that men and women experience friendship love differently – groups of mates for humorous relaxation for men as opposed to a single best friend for intimate chats with women – and that their parenting attachments differ – nurturance and challenge in dads as opposed to nurturance in mums. A recent review of the fMRI and PET studies which explore sex differences in the processing of emotion, encompassing thirty-two studies, found that the evidence is strong for women being more emotionally perceptive than men and being more reactive to emotional stimuli. This in part may be linked to the assertion that, anecdotally at least, women report feeling emotions such as love more intensely than men, who appear to have superior skills when it comes to the regulation of emotion. This split between the apparent rationality of men and emotionality of women leads to many heated arguments about the stereotyping of the genders, something we will address below, but this difference is reflected in the levels and location of neural activity when men and women are engaged in a task which asks them to consider romantic love.

Take thirty-two people who are romantically in love, sixteen men and sixteen women, and place them individually in an fMRI scanner. Show them a series of pictures of incidences of love from the classic couple holding hands in front of a sunset to carrying out the weekly supermarket shop. What would you see? Chinese psychologist Jie Yin and her team did just this to try and understand whether differences we believe we perceive in men's and women's approaches to love had any biological underpinning within the brain. Overall, what you would see is that, regardless of sex, the more romantic the scenario – sunsets rather than supermarkets – the more activation you see in the emotional centres of the brain; the romantic images elicit a more powerful emotional response in both men and women. This is reflected in the participants' scorings of the image where no sex differences are seen in how strongly the participants rated the emotional content of the image; their perceptions were the same. But if we look beyond these similarities, sex differences do emerge both within the emotional core of the brain and within the neocortex. The caudate nucleus and insula are more active in men, as are areas of the prefrontal and orbito-frontal cortex, both of which are engaged in social cognition, including mentalising. This would suggest that, for men, evaluating and considering romantic scenarios is a more effortful process than for women, where it is more instinctive.

However, when I shouted across the office to my husband and told him this conclusion, he found it, in his words, 'profoundly patronising', for it implies that men are less emotionally literate than we are. Suggesting that there are sex differences in the brain is a trigger for many people, for it implies that the stereotypes are true and deeply ingrained by millennia of evolution. It suggests that our desire to break out of our gendered roles is useless and we should all get back in our boxes. But the fact that we see these differences in the processing of romantic love cues in the brain does not mean that this is a result of evolution, nor does it mean that men do not *feel* love just as intensely as women. (We'll meet some evidence to support this assertion later on in this chapter.) Remember that when rating the pictures both men and women perceived the same intensity of emotion. We know from previous chapters that the human brain is highly plastic, particularly in our earliest years, and that our environment has a key role in moulding the structure and functionality of our brain. So it may be that our culture – which tells boys that they are the rational sex while women are at the mercy of their emotions – has caused our brain activity to appear as it does on the scanner screen. It is the case that love, as with many other emotions, is highly gendered. Indeed, take a group of boys and girls whose ages span both childhood and early adolescence – say, from six to ten years – and ask them to draw a picture of someone who is in love and this gender division becomes clear. A study by French psychologist Claire Brechet showed that as children mature, regardless of sex, then their use of graphic romantic indicators, such as hearts, increases as their understanding of romantic love grows. But if we consider the results by sex, while there is no difference between the drawings of boys and girls at the age of six, by the time they reach eight and ten the sex, and indeed gender, differences are writ large. While boys draw a strong, athletic man in love, girls' representations of female love represent a model of tender nurturance, usually ac-companied by a pretty and attractive dress. As our children grow, they become aware of the way society expects them to represent their love, and their ideas about love – the strong man and the vulnerable, emotional woman – become gendered.

To truly understand whether the differences we see in the brain are as a result of evolution or culture, we need to take our scanner around the world and explore the brains of those who live in cultures distinctly different to ours. Some of these cultures we will encounter in the next chapter. The closest parallel I can draw at present comes from my studies with gay fathers. In hetero-sexual couples, where mum is the primary carer, we see distinct differences in the peaks in brain activations between mum and dad. For mum, it is in the emotional core of the brain; for dad, in a close parallel to Jie's study of romantic love, the peak is in the neocortex. *But,* make the primary carer a gay dad – arguably a different parenting culture – and we see peaks in both areas of the brain, the mum bit and the dad bit. He has broken through the gender stereotypes which dictate dads provide and mums care. His brain is capable of encompassing *both* roles and many gay fathers celebrate the freedom their sexuality gives them to break down gendered parenting barriers. Maybe if we took our scanner to a society where men were allowed to be as open with their emotions as women, the sex differences would disappear. Or maybe not. When it comes to untangling the direction of causality – does the brain drive behaviour or behaviour the brain? – the web is as complex as ever, but in all probability what we have is a feedback mechanism which continues to impact our brain and our behaviour throughout our life course.

The Genetics of Singledom

Love to me is mutual respect and giving each other space to be your own individual. If somebody loves you, they respect you, they care for you, they are there for you and you in return. They let you be your own individual and live your own life, on your own terms, without judgement. Seeing somebody for who they are, not what they could be or should be. That is love. Acceptance is one of the hardest things in the world. **Flora**

To this point this chapter has been all about the genes and one gene in particular, OXTR. The evidence for a role for OXTR in our individual experiences of love is strong in part because its polymorphic nature means it has great influence but also because, due to the ease of sampling oxytocin, it is the most studied. As I mentioned in Chapter 2, the other neurochemicals which shape our love are a little trickier to access as they do not cross the blood–brain barrier, meaning that association studies between genes and neurochemistry are trickier to get off the ground. But when we go to the effort to do so, we get some interesting, and in this case highly newsworthy, results.

In 2014 the front pages of the daily press in the UK exploded with the headline 'They've found the gene that explains why you are single!'. Sharp intake of breath while the single people of Britain swiped rapidly through the pages on their phone to find out whether or not they were doomed to a life of single microwavable meals for one and talking to their house plants. As usual, the reality is not as exciting or as extreme as the headline. A group of scientists in China, based at Peking University in Beijing, had decided to explore whether, alongside the well-known input that personality, socioeconomic status, demographics and physical appearance have on the likelihood of being single, our genetics played a role. They chose the 5-HT1A gene, which has a single nucleotide polymorphism, rs6295, and codes for the serotonin 1A receptor. They recruited 579 Han Chinese students, 70 per cent of whom were female, and in addition to genotyping them for rs6295 asked them the very straightforward question 'Are you currently in love (romantically involved) with someone?' What a refreshingly pared-back approach. Their results showed that, in the first instance, there was no difference in their results regardless of biological sex, enabling them to pool all their data together. They then controlled for all the other factors that might explain why someone within their pool of participants could be single – for example, socioeconomic status, demographics, physical attributes and even how you were parented by your mum and dad – and ran their analysis again. What they found was that if you carry a G allele, so you are either GG or CG, you are significantly more likely to be single than if you are homozygous for C (CC). Genetics were playing a role. *But* like any other factor which influences how we love, this effect was *small*. 1.4 per cent of the variance could be explained by a person's version of 5-HT1A. So, contrary to the headlines, there is no gene *for* being single. There is simply a gene which may play a role, all other factors being equal, as long as the gene actually expresses itself in you, the phenotype. Unhappily single people of the world, rest easy.

However, the question remains: why does this gene in particular impact single status? The authors of the study, Jinting Liu, Pingyuan Gong and Xiaolin Zhou, point to previous studies which

suggest that 5-HT1A G carriers are less comfortable in close relationships, have more neurotic personalities and are more likely to
experience poor mental health, all of which can impact on your
likelihood of being able to maintain long-term relationships. I
have another idea to add to the mix, which is that they are simply
just not obsessed enough with their potential partner to bother to
make the effort to be in a relationship with them. Remember that
serotonin is implicated in obsessive compulsive disorder. Here,
low levels of serotonin drive obsessive behaviour and we know that
when you fall in love a drop in serotonin is hypothesised to drive
your obsession for your new love – lover, baby or friend. Maybe it
is simply the case that those who carry the G allele of the 5-HT1A
gene simply experience a greater serotonergic effect because the
density or affinity of their Serotonin 1A receptors are higher. As a
consequence, the obsessive elements of love, necessary to want to
maintain a connection to your lover and coordinate your life with
them, simply aren't there.

Love and Sexuality

*Love is giving someone permission to be themselves and them giving
you permission to be your most true self. To be comfortable around
someone. To stop performing. In another way, it is peeing with the
door open.* **James**

In this chapter we have considered how your biology – your genes,
your age, your ethnicity and your sex – can influence your individual experience of love. But there is one glaring omission still
to explore and that relates to your sexuality. At some point during
the Q and A sessions following my talks, the focus of the audience

will shift to the impact that sexuality has upon our experience of romantic love. Does our sex and the sex of our lover make any difference to our experience of love? Unfortunately, the majority of academic work on homosexual love, while claiming to explore the experience of love, is in fact about sex and how we differ in our sexual attraction to our fellow humans. However, what we do know is this. All love – platonic, romantic, parental – is underpinned by the same set of neurochemicals regardless of its type. And while the brain activity associated with parental love, platonic friendship love and romantic love do differ, this can be explained by the simple fact that maintaining friendships is cognitively more demanding than parental or romantic love and romantic love involves an element of sexual attraction, of lust, which, the last time I looked, is an aspect of your romantic experience regardless of your sexuality.

In 2010 evidence was produced to support this conclusion. Continuing his work on the neural bases of love, Semir Zeki, this time working with John Romaya, explored the brain patterns of romantic love in twenty-four individuals – six homosexual men, six homosexual women, six heterosexual men and six heterosexual women. As in his previous studies on romantic love, Semir placed his participants one at a time into an fMRI scanner and showed them pictures of their lover interspersed with pictures of a platonic friend of the same sex as their lover. What their results showed was that regardless of age of participant, length of relationships or intensity of reported love – as measured by the Passionate Love Scale – there were no differences in the areas or intensity of activation in the participants' brains regardless of their biological sex or sexual orientation. Indeed, the patterns they saw for all replicated the 'fingerprint' of romantic love they had first spotted on the scanner screen a decade before. Now it is the case that this is an unreplicated and very small-scale study and the participant group came from a wide age range and ethnic background. But in light of these findings, and in the absence of evidence to the contrary, I believe that the differences between individuals in their experience of romantic love that is due to culture, genetics or early life experience is greater than any difference between them due to their sexuality. The only exception comes when we consider the impact that a society's attitude to homosexuality has on an individual's experience of love. And that will be, in part, the subject of the next chapter.

CHAPTER SIX

PUBLIC

● ● ● ●

Legendary fifties crooner Frank Sinatra once famously likened the partnership between love and marriage to that between a horse and carriage, boldly claiming that one cannot exist without the other, and while he might have been over-egging this connection, it is undoubtedly the case that in every human society the love between two individuals is seen as a public issue. When you fall in love, you quickly become the recipient of torrents of opinion: on the character or appearance of your chosen mate; on the acceptability or otherwise of your love; on the socially prescribed length of time you should wait before any ritualistic celebration of your relationship can occur. Marry in haste, elope or defy your family by your choice and you will be left in no doubt as to the concerns or disapproval of your social group. Expand your viewpoint to look at love beyond the romantic and it is clear from our ongoing obsession with the lives of celebrity parents that your family is public property. Be seen to delegate too much parenting to the nanny or pronounce too vociferously about the benefits of attachment parenting and you will be met with a tidal wave of public ire from one sector or another.

It is very much the case that the love lives of others, the relationships they build with other people, are catnip for our brains. We have evolved to dedicate the vast majority of our cognitive brain power to observing, analysing and memorising the social lives of others. This was brought home to me during my masters at University College London when my course tutor for the evolution of cognition, Dr Cathy Key, set us a little quiz. The first part of the quiz was on basic geometry, the sort of thing in which we should all have been well versed as a consequence of our secondary schooling. The second part focused on the social lives of the characters in a well-known British soap, *EastEnders*, to coincide with the same time period in our early adolescence. Incredibly, while we struggled to carry out the most basic geometrical construction, all of us could recall in perfect clarity who had married whom, who had had a fight with whom and who was so and so's second cousin twice removed in this biweekly soap. We had perfect social recall. Why were we so brilliant at TV trivia when our mathematical ability left a lot to be desired? In part because the vast majority of our neocortex is given over to social cognition. Indeed, there is a direct positive relationship between the size of a species' neocortex and the size of their social group. And why do we dedicate so much brain power and energy to this particular topic? Because the stuff of our social network is the stuff of survival, as we know from the first chapter. Understanding the relations that everyone else has in our group means that we know who we have a chance with in the mating game, when an opportunity to rise through the hierarchy occurs, who might be plotting our downfall and who best to build a friendship with so their exemplary DIY skills can help us when our sink springs a leak or we need some shelves putting up. As a consequence, we dedicate a large proportion of our conscious brain to making sure we pay attention to, and remember, what everyone else is up to. In contrast, the ability to accurately construct a triangle with side lengths of 3, 4 and 5 centimetres is less critical to our survival and success. Who we and others choose to love – friends, family, lovers – *is* a public issue.

We split up when we were twenty-three and had been together for two years. We were apart for a year and then started seeing each other again and we were like 'What are other people going to say?' because my friends literally called her evil and her friends thought I was too clingy. We felt that everyone had a bad feeling about how things ended so we shouldn't tell anyone that we were back together. I think the main reason we hid it was the disapproval which we wanted to avoid. **Jake**

In this chapter I want to explore the norms and rituals that have evolved in human cultures around the experience of love; definitions of romantic and parental love; and what is acceptable, or even legal, in different societies. For love has broadly two dimensions – the biological and the social – and each has an influence on how we perceive and experience love. However, until the early nineties a question mark remained over whether or not 'romantic' love – love with a reproductive partner – was even a human universal. An outdated Eurocentric ethnographic view was that without the published literature and art which had helped us conceive of romantic love in the West, this was simply an experience denied to those from non-Western culture. From our position, grounded as it is in a science of love which acknowledges love as a physiological and neurological universal, this would appear to be a ludicrous position and denies the rich oral storytelling traditions of these cultures, but it took a ground-breaking ethnographic study by social anthropologists William Jankowiak and Edward Fischer, based upon observation rather than assumption, to confirm the presence of romantic love in all cultures.

In their 1992 study of 166 countries across North and South America, the coastal Mediterranean, northern Sahara, east Eurasia and the Pacific Islands, they found that earlier conclusions that romantic love did not exist in these societies were due to spurious assumptions and a clear absence of curiosity by the ethnographers involved rather than any absence of evidence. They had simply failed to ask the questions, firm in the belief that romantic love could not be present in these cultures. Regardless of your background, love for a sexual partner, whether or not they are your socially proscribed spouse or a lover, is a constant in all societies.

Love in Action

I suppose I am or can be a bit embarrassed about my polyamorous relationship with Kate. Kate is out completely with her family . . . I am less comfortable with that and am not out in that sense with my sisters. I have two sisters who are quite a bit older than me who I just don't think would understand it. **Jeremy**

It is quite clear that early ethnographers were wrong and that love in the romantic context is a human universal, but if we were to be benevolent and try to understand their misunderstanding, it could be that they were confused by the fact that, while romantic love is felt by all, *how* we display that love to the wider world is culturally constrained. Indeed, of all the feelings – anger, happiness, sadness, fear, pride and love – love is the one which is the most culturally specific. While cross-culturally we are able to use the facial expressions, body movements and voice tone of other cultures to correctly interpret these other affective states, this is more tricky when it comes to love.

In 2017 a team of American psychologists led by Carolyn Parkinson from UCLA travelled to the remote Cambodian region of Ratanakiri to explore whether its residents, the Kreung hill tribe, could correctly interpret these six states when displayed by an American 'poser' relying solely on body movements. This particular Kreung group live in the village of L'ak, which is cut off from the outside world for much of the year. They speak a dialect which is not understood by the majority Khmer-speaking population and, because they are self-sufficient, rarely have contact with anyone beyond the group. As a consequence, it is highly unlikely that their perception of love has been influenced by access to western culture. The American poser was asked to interpret the six feelings in body movements and these, having been recorded, were shown to twenty-six members of the Kreung community, eleven of whom were female. It was tricky to know how old the participants were as the Kreung do not document age, but all were from late adolescence to older adulthood. The group were shown fifteen-second clips of the poser performing the body movements associated with the given feeling and asked, without the aid of prompt words, to pinpoint the feeling they were observing.

So for anger, the poser displayed a range of movements such as a rapid pace and irregular, erratic and forceful movements. In 100 per cent of cases the Kreung were able to recognise this display as anger. For happiness, the poser kept a moderate pace, their limbs and shoulders were relaxed and they swung their arms from side to side. In this case, twenty-four of the twenty-six Kreung participants correctly recognised this as happiness. But when it came to love, their run of good performances came to an abrupt halt. The poser carried out movements including relaxed arms and shoulders, with arms swinging at the sides, hands over heart and twirling around, but in only three out of twenty-six cases did the participant correctly identify the feeling as love. They were more likely to attribute the behaviour to happiness and, less understandably, anger and sadness. In contrast, American raters identified the performance as love in twenty-four out of twenty-seven cases. It was clear that for love the traditions and cultures in which we are raised have a profound impact on how we display our love.

When I first got together with my wife, it was hard because I was anticipating negative reactions that never came from her side and our friends. And I think I was always expecting something to go wrong. I think the first few years I felt very vulnerable because I held myself so secret for such a long time I just thought how vulnerable it was for people around us to know about us and be fine about it. It made me feel stressed. It took a while to get used to it. **Lily**

And as our behaviour when in love is culturally proscribed, so are our definitions of love as well. Compared to the Greeks, who famously have seven words to describe different forms of love, and the Arabic language where terms for love include *ishq* – the love that entwines two people together; *hayam* – the love that wanders the earth; *teeh* – love in which you lose yourself; and *wlaah* – the love that carries sorrow within it, we have the ludicrously inadequate single word – love – to encompass all the feelings, experiences and relationships of love. And when we ask people what they find attractive, while all adhere to the basic evolutionary drivers of fertility for women and resources for men, we do see cultural variations.

One of the most fascinating studies was carried out on mate preferences in China. It is well known that China has undergone a dramatic shift in its economy and culture over the last thirty-five years, from strict communism to a version of communist capitalism which is increasingly Western-influenced. This has proved to be the perfect testing ground to explore how shifts in culture impact our love experience. Using two populations, one from 2008 and the other from 1983, psychologists Lei Chang, Yan Wang, Todd Shackelford and David Buss explored changing attitudes to desirable traits in potential mates. The 2008 group were 1060 individuals – 475 male and 585 female – who were recruited from a diverse number of locations including businesses, factories and universities. The older population were data which David Buss had collected as part of his ground-breaking study into love across thirty-seven cultures – a study which played a large part in putting the erroneous belief that romantic love was purely a Western experience to bed. This Chinese population consisted of 265 men and 535 women.

What Buss and his team found was that while in 1983 being religious was the least important attribute, by the 2000s this had risen by two rankings – perhaps reflecting the increased tolerance for religion within the state – and being of similar religious belief was a key factor for both sexes in 2008. Being a college graduate, with good earning capacity and of dependable character, were also significantly more desirable as attributes in the younger population. In terms of what had dropped in the rankings, while both sexes believed that being a virgin was practically indispensable in 1983, by 2008 it was well down the rankings. Why these changes? The shift in economic model which has led to a stratification in wealth in China is well documented, and this stratification means that, particularly for women, the consideration of economic potential is an important one. And the increasingly relaxed attitude to premarital sex is one which has been imported from the West as China opens its doors to more and more Western influence.

The Meaning of Love

All of my close friends and immediate family know I am aromantic. I didn't tell my family for a long time because I didn't know how they would react really. I don't tell my extended family, I don't tell my colleagues because I know the chances are they are not going to understand or accept it. **Sara**

Beyond the cultural influence on attraction, where we live has a significant impact on how we define love and this is in part influenced by how free we are to express that love. In his book *Romantic Love in Cultural Contexts*, anthropologist Victor Karandashev reviewed the research from around the world to draw conclusions about how different cultures interpret, and make rules about, love. For example, in Arab Muslim culture, to be in love as a man is to risk being out of control. Indeed, to be in love is the source of all evil, as a woman can bewitch a man and shift the balance of power from the patriarchy to the matriarchy. In Morocco, while a man is free to choose his partner based on love, a woman is not. Love is seen as having magical power, embodied in a spirit known as Lalla Aisha who may visit a man in his dreams and make him impotent. Men are seen to be possessed by love, making them irrational and impetuous. It is for the woman to bring the rational balance to the relationship and she is socialised from a young age to base her decision on practicalities; will the man make a stable marriage partner or tie the woman to a good family? She is actively encouraged to think with her head, not her heart, and as a consequence women tend to report feeling passionate love less intensely (whether they do or not in reality is another matter). Indeed, to admit the possibility of romantic love to anyone other than her closest female friend would bring her purity into question. In Bedouin culture two forms of love are recognised; *ilhub,* which describes passionate love and is likened to a sickness, a life-threatening condition, one of pursuit and desire; and *ralya,* which means 'precious' and 'dear' and describes companionate love towards family members, one's spouse and one's friends. Why this distinction between the two; one seemingly dangerous and uncontrollable, the other calm and reassuring? Because in Bedouin culture marriage is a matter for the patriarchy and not the individual, so any hint of passion is to be discouraged as it risks leading to a partnership which cannot be controlled.

Cultural ideas about love can also impact the love a parent may openly display to their child. In Tamil culture a mother's love should be hidden – it should display *Adakkam* (containment). Boundless love is seen as harmful to both the child and the mother, and so acts of affection are frowned upon. She will regularly shower other people's children with affectionate words and actions, but she must not even gaze affectionately at her child when they are asleep, as this risks the evil eye. In Tamil culture, to love your child is to deliberately withhold affection, to experience self-sacrifice as you work against your natural instinct to shout the love you have for your child from the rooftops. Finally, if we were to return to the strictly communist China of the nineties, then, in contrast to the West where love is associated with happiness, affection, passion and caring, definitions of love are couched in terms of sadness, unrequited love, infatuation, nostalgia and sorrow. It would appear that during this time the Chinese population felt love to be out of reach and hence a source of sadness.

The riskiest thing I have done in the name of love is to be poly. Because it is not 'normal' in the sense that it is not a normative behaviour and it makes you worried that if other people know and don't like it they will reject you because you are not conforming to society's norms. **Kate**

As we have seen with China, as the cultural barriers between countries break down – as a consequence of globalisation – definitions of love tend to shift more towards those found in the West. However, in a recent study a group from Russia explored representations of love from three, still ethnographically diverse, populations: Brazilians, Russians and Central Africans. Their aim was to understand cultural similarities and differences in the perception of love, and to inform the mediation of conflict in intercultural relationships which they believe to be in part due to a mismatch in the definition of love. For our purposes, their study is a fascinating insight into how our perception and experience of love can be strongly influenced by where we reside. Tatiana Pilishvili and Eugenie Koyanongo recruited fifty people from each population – twenty-five men and twenty-five women. In the first instance, participants were asked to name the first three words which came to mind when they heard the word 'love'. Secondly, they were asked to complete The Classical Ideas of Love: Acceptance and Distancing measure, which presents participants with twenty-six sayings or writings about love by individuals such as Shakespeare, Voltaire and Tolstoy, and asks them to indicate to what extent their ideas about love agree or disagree with the statements.

Their results were fascinating. For all the cultures, love is a source of joy and is the manifestation of all that is special and good in a person. But beyond this, other aspects of love were culture specific. So for Brazilians the most frequently associated word to describe love is 'honesty', and they perceive love as being grounded in feelings, morality and family. In contrast, the most frequently mentioned word for Russians was 'suffering', and love to them is associated with trust, self-sacrifice, hope and, in common with Brazilians, emotions and family. Finally, for the Central Africans love is embodied in 'tenderness', and their conception of love is grounded in the spiritual. God is love and as the extension of God on earth their experience of love is characterised by purity and action – to serve and support others, to trust, to give oneself to another completely, to be attached. The Brazilian experience of love is passionate, honest and intuitive. For Russians, love is about continuation of the family, and while it has elements of romance it is also cognisant of the potential difficulties love can bring. While for Central Africans love is divine and ennobling – they experience elevation through their support of others.

The Spectrum of Sociality

When I started one of my first jobs in the UK, I hadn't come out because I just assumed you don't talk about these things when you start a job. I had worked there for six months when I got a partner. Someone found out and they made a lot of comments and I had to explain myself to HR as apparently I had lied, which I hadn't, I really hadn't. She said she felt uncomfortable with me. From then on I came out at job interviews, but subtly, so I never experienced that again. **Lily**

All of the societies which made up Tatiana and Eugenie's study, and indeed Chinese and many Arab Muslim communities, are described as collectivist. This means that the individual's needs are subordinated to those of the group. Love is often described in terms of the wider group and how love can bring benefits to others; the emphasis is on companionate rather than passionate love. In contrast, in the West, where most countries are individualist, love is often described as a personal experience and feeling with little reference to the good our love may bring to the wider group. This dichotomy between collectivist and individualist societies is one of the elements of our social environment that influences our expectations about what love is, how we perceive love and how we behave and feel when in it. In collectivist societies passionate love is often seen as a negative, a threat because it leads the individual to do what is best for them, to follow their heart, in direct opposition to what is seen as being best for the society, which might be marrying someone from a particular class, ethnic group or religious persuasion to maintain the status quo. In the individualist West this is seen as outrageously controlling; love is about freedom and is the ultimate expression of the individual. Indeed, to pair up in the absence of passion is seen as a serious risk for the long-term happiness of the people involved; to not experience romantic love is to live only half a life.

When I got back with my girlfriend, I had one friend who sent me a text message saying 'I have warned you enough times about this and while you are being stupid I don't want to talk to you. Get in touch once she is out of your life again.' **Jake**

Beyond the collectivist–individualist spectrum, other factors come into play in influencing how open a culture is to the idea of romantic love – as opposed to companionate love – being the basis for long-term relationships. In turn, this influences how acceptable it is for romantic, passionate love to be openly displayed and talked about within the society. For where society dictates that marriage and love do not align, love is seen as dangerous both to the individual and wider society and is actively suppressed. In their 2016 paper 'Romantic love and family organisation: A case for romantic love being a biosocial universal', evolutionary psychologists Victor de Munck, Andre Korortayev and Jennifer McGreevey argued that female status and the nature of family structure were important variables when predicting whether romantic love was visible, permissible and deemed valuable within a society. Based upon their analysis of the data from seventy-four societies crossing all continents, they argued that where female status is high – that is, at least equal to that of men – and the norm is for nuclear families, romantic love is the basis of marriage and is widely accepted and valued within the society. In contrast, where women have low status and the family includes extended kin, then romantic love is actively suppressed and is deemed to be unacceptable, even dangerous, as the basis of marriage. Why is this? Romantic love is based on the premise that its members are unique and that the love signifies the meeting of equals. Where women are seen as significantly inferior to men, this cannot be the case. Likewise, where the extended family is the norm the need for the intimacy between the couple which sits at the centre of the nuclear family – co-sleeping, eating and socialising – is reduced, as is the need for a powerful love to bond man to woman as the two sole providers of childcare. Where extended family do a significant amount of childcare, then romantic love is less necessary to bond mother to father as, should the father leave, the pregnant mother has many others to turn to for support in raising her child.

Further, where the extended family is the norm, family members feel they have a legitimate role in choosing a spouse for their child if they are to invest a considerable amount of time and resources into raising the offspring of any union. In these circumstances, romantic love is suppressed. This is not to say, as argued by early ethnographers, that a concept of romantic love does not exist in these societies – it is certainly evident when subjects describe extra-marital relationships – but only that it is not acceptable as a basis for marriage, and as such romantic love is not an acceptable topic of conversation, nor an open driver of behaviour, in these societies.

The Rules of Love

When I was in my early twenties, I started a new job. There was a guy sat a few desks away who was quite a lot of fun. We started bantering on the internal mail . . . One thing led to another and we decided we would start going out, but it was clear we shouldn't say anything at work. We had to keep it quiet, which meant behaving differently in the office from when we were at home. It was trying to remember who knew and who didn't and keeping your behaviour straight for the right people. It was challenging. We did not intend to come clean. We ended up eating some poisoned yoghurt and we ended up on the front page of the Daily Mirror *and the* News at Ten! *Because it was such a bizarre outing, no one ultimately minded about the relationship because we had ended up in such a ridiculous situation.* **June**

It is clear that where we were born, raised and live has a significant impact on how we perceive, define and express love. All societies have rules about love which shift over time as political, religious and economic environments change. Even here in the West, where the perception might be that we have freedom to love who we like, in the way we like, there are still rules about what love should look like. In her book *Why Love Hurts* sociologist Eva Illouz describes the rules of love as dictated in the time of the British novelist Jane Austen, renowned for her depiction of the minutiae of love, marriage and society among the middle and upper classes in eighteenth-century England. In the first instance, for a woman love was not an individual affair but a collective one. Rules denying the acceptability of a young woman meeting alone with a gentleman meant that family were able to observe and scrutinise the man to assess whether his intentions were matched by appropriate displays of behaviour and emotion. Courtship ran to a strict timetable – detours from which would call into question the gentleman's intentions and true feelings. First moves were only permissible by the woman, who would invite a potential suitor 'to call'. Couples first spoke, then walked out together – chaperone following discreetly behind – and then, and only when mutual 'strength of feeling' had been confirmed, the couple were allowed to meet alone. Eva refers to this ritual as 'a regime of performativity of emotions' – the rules dictate that the performance of the ritualised behaviours of courtship induce the emotion in the couple rather than the emotion necessarily being a prerequisite.

To the modern British ear, this sounds suffocatingly restrictive and a potential waste of time and energy – why go through the motions of the ritual if love is not guaranteed at the end? But despite our protestations, our experience of love in the UK is still bound by rules. Rather than being based in ritual, they are based in, as Eva coins it, the authenticity of our emotions. Emotion – love – must be the driver of action and we make decisions regarding our future based upon the intensity of these feelings. We may canvass the viewpoints of others, but ultimately the needs and beliefs of the individual are paramount. We analyse the rightness of our love based upon how we feel, and question those whose relationships appear to be motivated by factors other than love, for they are deviating from the socially accepted rule that romantic relationships must begin and end with love. The derogatory term 'gold digger' sums up our disapproval if anyone dares to marry for money rather than love. To remain within the bounds of what is socially acceptable we must protest our love even if it is not the dominant driver in our relationship. Contrast this to those societies where to marry for anything less than money and status – to marry for love, say – would be met with incredulity.

So, love is bound by societal rules. If we look at the nature of these rules, while some relate to the appropriate ways to display love or the value or necessity for love, there are many rules which relate to demographics; age, sex, ethnicity and class. The marriage rules of India are well known as they relate to caste and religion. While the literature and films of India are wildly romantic, what is striking is that they often portray romantic love as something to be hidden from the family and wider society – the characters are often pursuing doomed relationships across religious and class divides. They are depicted as meeting in magical, private places where they can love away from the gaze of those who might stop them. The gap between their fantasy and the reality of their personal destiny – where a partner vetted by family and religious leaders awaits – is often the main theme. Interestingly, despite the encroachment of the West into many other areas of Indian life, the rules relating to interfaith and inter-class marriages remain strong. This may be in part because of the increasing stratification of society based upon wealth, which has resulted in those who frown the most on love matches being the young upwardly mobile; at the very most, marriage must be with someone of similar if not higher class to ensure progression. And the resurgence of religious divides between Hindus and Muslims has made the need to assert one's identity through faith more important. To love someone from across the divide is a betrayal of one's culture and identity.

Love Across the Divide

My mum has an issue with my relationship with a Syrian refugee. It's the status she doesn't like, not race or religion. It's about trust or thinking he is not good enough. I am still living my relationship but it does mean I am not sharing things on social media for example. I have told her about my relationship and she was not happy to say the least. I was shut down in the most horrific ways I could imagine . . . the conversation went on for 8 hours and she was throwing ridiculous things in my face. We haven't spoken about it since. I find it sad for him and my mum and there is a big part of my life she is not part of. I still speak to her about life and what we are doing but with this big part missing. **Charlotte**

The need to preserve religious and cultural identity in the face of threat, and the impact this has on the permissibility or not of interfaith relationships, is no more striking than in the attitudes of young people to love across the ethnic divide in Israel and the West Bank. In their 2016 study entitled ' "Sadly Not All Love Affairs Are Meant To Be . . ." Attitudes Towards Interfaith Relationships in a Conflict Zone', a team of psychologists and conflict resolution specialists, led by Siham Yahya, explored attitudes

amongst young Palestinian and Israeli adults to interfaith relationships. They carried out semi-structured interviews with eighteen
participants – fourteen female and four male – seven of whom
were Jewish, six Christian and five Muslim. All were citizens of
Israel, but seven identified as Jewish and twelve as Palestinian.
Beyond the rules prohibiting interfaith marriages, which are present to a greater or lesser extent in all three religions, because Israel
and Palestine are at war it is also a federal offence – punishable by
losing one's citizenship or even prison – to marry someone of a
different faith.

Four main themes emerged from the interviews. The first was
that interfaith relationships were immoral due to their disregard
for religious rules and as a consequence the young people would
avoid such relationships. Here's one quote from the study:

*I never imagined that one day I will fall in love with someone;
that is, I never gave myself the opportunity to fall in love with
or think of someone not from my faith or culture. I am all for
sacrificing for love, but if I was ever in that situation, I would go
to the high religious people and inquire, maybe in earlier stages,
but sadly not all love affairs are meant to be.* **Rania**, a Muslim
Palestinian woman

Rania's beliefs were echoed by others in the study who believed
that they had ultimate control over who they would fall in love
with. Their head would definitely rule their heart.

The second theme centred on the role of the importance of
familial approval. Many were loath to defy their family due to the
risk of losing contact with them or, for one participant, the belief
that his father would commit suicide were he to fall in love with
someone of a different faith. The third theme brought home the

role of the conflict in reinforcing beliefs about interfaith relationships. Many felt there was strong societal pressure not to betray one's ethnic and religious identity – to identify with the 'other side' – and some went so far as to say, in the current climate, that doing so was treason. Many pointed to the fact that alongside physical segregation, which meant that for many meeting someone of another faith or ethnicity was unlikely, there was strong cultural segregation. Children were taught from early on the importance of marrying within one's faith. The conflict played a major factor in young people's thoughts about love across the divide with one questioning how 'you can create love in a place that is so full of hatred and politics and opinions and extreme ideologies'. The final theme looked at the importance to these young people of maintaining their cultural identity going forward, which meant that not only would they actively avoid falling in love with someone of different faith but that they would teach their children to do the same. They questioned how it would be possible to maintain their beliefs and culture in their children if they had to compromise with a partner who held different beliefs.

Chaos and Control

I grew up in rural France in a very heteronormative, homophobic place . . . I grew up thinking I was very strange and I felt different. Everything around me made me feel like I had to pretend to be something I wasn't and then I grew up and thought I think I may be queer. But my entire environment was peers saying 'gay people need to be exterminated', my parents would say 'you can never trust to leave children with gay people' . . . so I hid who I was for most of my childhood and teenage years. **Lily**

The social, religious and cultural rules of love are ultimately about control. They are about making sure that cultural identities are maintained – the importance of defining self against other which is critical to all human groups – that power stays within certain groups or families, that wealth is controlled by the few rather than the many and that society is, to a greater extent, stable and predictable. Love threatens this status quo because it is so powerful and so unknowable. As we will see in Chapter 10, it motivates people to undergo the most amazing trials and tests, to take on nigh-impossible hurdles to ensure they can be with the one they love and that makes those in power scared. The fear is that to fall in love is to give your devotion, your obedience to the one you love, meaning that they, rather than the state, are your ultimate master. This quite clearly shows a misunderstanding of love – it is not a form of madness which makes us lose all rationality. However, such is our fear that we put in place strict rules to make sure everyone stays well within their box. From an evolutionary point of view, such rules make sense. Like all social animals, we exist within a hierarchy which dictates who we mate with, how much power we have, our access to resources and how successful our offspring may be. Everyone has their place. The rules of love have developed because by understanding and conforming to them – compliance sometimes being 'encouraged' via threat and force – we expend less cognitive energy on monitoring our fellow group members. What would become of our ordered hierarchy if we were allowed to lust after, procreate with or befriend whoever we wanted, whenever we wanted? Even our enormously adept social brain might struggle to cope with that one.

But this suggests a rather negative view on the rules of love. Not all rules confine. Some liberate and provide structure and guidance as we try to navigate what can often be the scarily complex world of dating and love. Others mean that we are able to communicate our intentions clearly and comprehend those of others. This is the argument put forward by sociologist Eva Illouz, who believes that the rules or rituals of love help us to deal with the uncertainty that often accompanies love – what should I do next, how should I feel, what is the other person trying to say? And it is certainly the case that for a significant number of the people who I talk to about the science of love, it is certainty they are seeking. They find today's world of dating – with its myriad apps and apparently endless choice of potential partners – confusing and stressful. They hope that science can answer their questions and give them some idea of what their feelings mean and how they should behave. No pressure then.

Prohibited Love

Here are some facts from the UK:

Only 46 per cent of LGB (lesbian, gay and bisexual) people feel able to be open about their sexuality to their family.

A third of LGB people of faith aren't open with anyone in their faith community about their sexuality.

35 per cent of LGBT people have hidden their identity at work because of fear of discrimination. Almost 1 in 5 have experienced negative comments or behaviour from co-workers.

At university 28 per cent of LGBT students were excluded by other students because of their identity.

The number of LGBT people who have experienced a hate crime in any given year has risen by 78 per cent since 2013.
'LGBT in Britain' – Stonewall, 2018

If these were (are) the facts of your life, how comfortable would you (do you) feel about being open about your romantic love life? Homosexuality is still illegal in seventy-two countries, mostly in Africa and the Near and Middle East. In these countries, governments who should be there to protect their citizens and maintain order often turn a blind eye to attacks on LGBTQ+ people. Amnesty International, the human-rights charity which campaigns against injustice and inequality worldwide, reports state-sponsored campaigns in Chechnya which see the targeting of gay men, some of whom are abducted, tortured and even killed. In Bangladesh machetes have been used to hack LGBTQ+ activists to death with police showing little interest in bringing the perpetrators to justice. In nine countries – including Pakistan, Qatar, the UAE, Nigeria, Somalia and Iran – having sex with someone of the same sex is punishable by the death penalty, and while this punishment isn't always enforced, the fact of its existence means that LGBTQ+ people live in fear of harassment and blackmail. For most, being open about who they love is a risk too far.

And even where homosexuality isn't illegal, considerable barriers can be put in the way of LGBTQ+ people. In Russia the 2013 legislation which outlawed the public promotion of homosexuality, particularly to children, has meant that support for young LGBTQ+ people is practically non-existent and President Putin, while claiming not to be prejudiced against gay people, has stated that homosexuality is out of step with Russian traditions and culture, meaning that same-sex marriage is likely to remain outlawed, and gay people may not adopt. These are not the behaviours of a country where open displays of gay love are tolerated. Even in the West, where we have come a long way in accepting gay love (twenty-seven countries have legalised same-sex marriage so far), recent figures from the UK show that one in five LGBT people has experienced physical or verbal attack due to their sexuality, indicating that we still have a long way to go before these forms of love are truly publicly acceptable.

Coming out to my family was very painful and I am now estranged. For a long time what I didn't know was that my mother, who knew I was gay, was hiding it from other people. She was ashamed. I met my now wife and we got married and it was a really awful process. They met my wife at my wedding. They refused to see her before and the compromise I had to make for my parents to come to my wedding was to not tell anyone else in our family. Even though they came and I thought it was the end of it, it wasn't. She was OK if we were indoors, but if she walked next to my wife and I in public she would walk ten metres back. **Lily**

For those who identify as LGBTQ+, this can mean that any public displays of love can lead at the very least to the risk of rejection by family and the wider community, and at the other extreme to prison or even death. In these circumstances it is critical to avoid your love becoming a matter of public knowledge. There is no sharing your joy or heartache with your family, friends or colleagues, and the chances of you experiencing that other form of love, parenthood, are slim. While your love is as powerful and as valuable as anyone else's, your experience of love is likely to be markedly different because of the need to deny publicly who you are. And even where being gay is more accepted, those who identify as gay or bisexual often have to consciously consider when and how to be open about their orientation and their romantic relationships. In their 2003 paper exploring the experiences of gay, lesbian and bisexual adolescents, American psychologists Jon Lasser and Deborah Tharinger reported on the outcomes of interviews with twenty LGB youths. The central theme they identified was one of visibility management; the process by which the interviewees constantly reassessed the decisions they were taking about being open about their sexuality. Should they disclose their sexuality? If so, who do they declare it to and how? And how do changes in the environment impact these decisions? Decisions regarding the degree to be open could change on a moment to moment basis. This level of constant conscious awareness of the acceptability or not of openly expressing our love is something that those who are heterosexual are very rarely ever going to have to consider.

This chapter has been about how our culture impacts both how we perceive and habitually display love and the 'love' rules that all societies impose on their people in the hope that the status quo can be maintained. We have learned that uniquely among the affective states, our displays of love are culturally specific. That how we perceive and define love is influenced by our culture, although creeping globalisation means that Western ideas about love are increasingly prevalent. We understand that all societies have rules about love which may constrain or free us, but are ultimately about control. And that the illegality of some forms of love means that, for some of us, being able to be open about our love is but a dream. But there is one social rule we haven't considered yet, and that relates to the concept of monogamy, the socially acceptable face of romantic love in the majority of the world's countries. Can romantic love really only exist within monogamous relationships? That is the topic of the next chapter.

CHAPTER SEVEN

EXCLUSIVE

● ● ● ●

*'Wilt thou have this woman to thy wedded wife, to
live together after God's ordinance in the holy estate of
Matrimony? Wilt thou love her, comfort her, honour, and
keep her, in sickness and in health; and, forsaking all other,
keep thee only unto her, so long as ye both shall live?'*
from The Marriage Service, *Book of Common Prayer* (1662)

Traditionally, Western concepts of love have aligned themselves with the idea of exclusivity, with the search for and capture of our one true love. Indeed, enter the jungle that is the New York dating scene and a key turning-point in your relationship will occur when you have the 'conversation', which signals your denial of all others and your couple exclusivity. Find love online and the arrival of exclusivity is heralded by the mutual ritualistic removal of Tinder from your devices. And, as every child knows, the fairytale ends with the man and the woman marrying and living together 'happily ever after' – there are never any additional partners waiting in the wings. But, the question arises, did human romantic love evolve to be monogamous or, as evidenced by the 25 per cent of American men and 15 per cent of American women who have affairs, and the 5 per cent of the American population who admit to a polyamorous relationship, is our obsession with 'the one' merely a social construct? An excellent bit of propaganda dreamed up eons ago by those who wished to tame romantic love?

In this chapter we are going to interrogate the idea that romantic love equates to exclusivity. That once we have found 'the one' (there's a bit of romantic propaganda if ever I heard it) we resist all others and give all of our love just to them. We are going to explore where these ideas come from and speak to those – the polyamorous – who have chosen to take a different path in their love lives. We'll hear their real-life experiences and contrast them with the ideas that society has about this identity which generally equate it with the division of love, immorality and promiscuity. And of course science will get into the picture as both monogamy and polyamory get placed in the scanner. At the other end of the spectrum are the little-heard-from aromantics, who claim to not experience romantic love at all. Their voices and experiences will close this chapter.

The Road to Monogamy

Monogamy is the majority practice in western societies and is sanctioned as the acceptable face of romantic love by civil and religious laws which outlaw both adultery and bigamy. Adultery is grounds for divorce in many countries, including the UK, where divorce proceedings are still based on fault, while in South Korea 5500 people were successfully prosecuted for cheating on their partner between 2008 and 2015, when adultery was decriminalised. And in some Muslim countries where Sharia law is upheld, stoning is the punishment for sex outside marriage. Why are we so scared by non-exclusivity? The answer lies in the fear that explains what zoologists call 'mate guarding' in primates and leads to jealousy in humans: that you will have to share the resources your partner brings to your relationship with others. So as the male hamadryas baboon fiercely guards his fertile female mate, so human males, in particular, enter exclusive relationships with the female object of their affection. For men, this is particularly key due to the fact that human females, rarely in the primate kingdom, exhibit concealed ovulation, which means, unlike the bright red bottoms of many a fertile non-human primate female, the tell-tale signs of ovulation are hidden. Because of this, to ensure that he is the male to mate with the female when she is most fertile, and therefore guarantee that any future offspring of the union are his to invest in, human males, unlike the male hamadryas baboon who only has to stick around for a matter of days, must 'guard' their female permanently. And for women, the exclusivity of monogamy is important because, while she knows any children are hers, she needs to make sure that all the male's resources – food, protection, education – are funnelled towards her offspring and their survival.

And because of this need for exclusivity, some of the most extreme and insulting language in human culture is linked to sexual infidelity. Additionally, if you try and write a list of these words it is likely that the list you will come up with for an adulterous woman is longer than that for an adulterous man. In English we have 'whore', 'slut', 'ho', 'trollop' and 'tramp', to name but a few. For a man we are limited to the Shakespearean 'cuckold', in this case the victim of adultery – which is hardly going to cause anyone's maiden aunt to faint any time soon. The reason for this? An unfaithful woman risks removing her partner's opportunity to father a child for a good nine months at least – her womb being otherwise occupied by the child of her lover – whereas for a wife, despite her husband's errant ways, the opportunity remains for her to carry her husband's child even though the risk exists that she may have to share some of his valuable investment with the children of another union. Beyond this, as we have seen from Chapter 6, the institution of marriage developed as a way to limit the havoc that unbridled love could wreak on society and make sure that wealth and power remained in the hands of a privileged few. These 'few' were most often led by men as a consequence of the overwhelming presence of patriarchy in nearly all human societies. As a result, women are blamed more, demonised more, should they stray. Take this further and make marriage a sacred institution and you make monogamy – and the breaching of it – a moral as well as legal issue.

We seem to have taken this message on board pretty comprehensively, although we have made it more palatable over time with a bit of romance. Think of the language we use about the quest for a relationship or to describe the love we have for our lover. Single people often yearn to find 'the one', the person who is on the verge of taking the next step with their lover towards lifelong commitment questions whether this person is their real destiny, and lovers often describe themselves as 'soulmates', which implies the meeting of two unique entities who can only be meant for each other (the concept of fate is often introduced at this stage too). Our love is meant not for the many but for 'the one'. The view is that romantic love is a precious commodity, the sharing of which diminishes its value to the object of our affection. This is known as the zero-sum view of romantic relationships and as a recent study showed, it is a belief that is deeply ingrained in our society.

The Zero-sum Game of Love

In their 2017 paper 'Wanting the Whole Loaf', American psychologists Tyler Bureigh, Alicia Rube and Daniel Meegan explored the public's attitude towards love in monogamous and polyamorous – that's romantic love with more than one person – relationships. They recruited 136 American participants, fifty-nine women, seventy-six men and one genderqueer, the vast majority of whom identified as monogamous. They presented them with a series of vignettes describing the relationship history of a couple called Dan and Susie who had been dating for two years. In one set of vignettes, Dan and Susie were polyamorous – their practice involved having more than one romantic partner – and in the other they were monogamous. In both relationship scenarios the scene was set with a vignette which described Susie and Dan and explained that they had agreed, in the polyamorous scenario, to have an open relationship (although neither had acted on this so far) and, in the monogamous scenario, to only have a sexual and emotional relationship with each other. After reading the establishing vignettes, participants were asked to rate how much Susie loved Dan, the quality of their relationship and the trustworthiness of Susie.

They were then asked to read a second vignette about the third year of Susie and Dan's relationship. In the polyamorous scenario Susie was still with Dan but had met a man, Oliver, with whom she was also having a sexual and loving relationship. In contrast, Dan had not met anyone who he had strong feelings for. In the monogamous scenario, Dan and Susie had stuck to their agreement and not had any other relationships. After reading these second vignettes, participants were asked to rate again how much Susie loved Dan, whether or not their relationship would be satisfactory in the long term, how likely it would be that Susie would be in a relationship with Dan after five years and, finally, how trustworthy they thought Susie was now.

The results were clear. By having a relationship with both Oliver and Dan, each man got less of Susie's love. So, while in the monogamous scenario Susie's love for Dan had in fact increased, in the polyamorous scenario it had significantly decreased; Oliver was now getting a portion of the love cake which had previously been Dan's. A clear example of zero-sum thinking. Further, the participants had evaluated both the levels of satisfaction and longevity of Dan and Susie's relationship and Susie's trustworthiness negatively, whereas these were deemed to be positive in the monogamy scenario. Participants saw love as a zero-sum game and because of this they viewed polyamorous love as less powerful for the individuals involved and more likely to be unhappy and at risk of ending. A further study exploring why people held zero-sum beliefs about romantic love confirmed that these were based on the two factors of entitlement – a lover should be entitled to all of their lover's love – and scarcity – love is finite and you can only love one person fully. A combination of social conditioning (entitlement) and evolution (scarcity).

A Polyamory Primer

I feel that polyamory is defined by the ability and the freeness of the love . . . every day my partners wake up and they choose to love me. Not because they don't have other options. Not because they are not exposed to other people . . . but they choose me because we are a good fit and because we enhance each other's lives in such a way that cannot be replaced by another person. And I really love that. I feel it is a very pure way to have that connection. This connection is not governed or protected by rules, it just is because it is so strong and so positive that even when we can choose anything we want we still choose each other. **Jo**

Romantic love would appear to be the exception when it comes to love. We can love many children, friends, pets, and gods, but we can only ever feel the excited flutterings of romantic love for one person at a time. Romantic love is an outlier. But is this truly the case? Speak to those who identify as polyamorous, such as my interviewee Jo, and they argue that this is quite clearly not the reality. Polyamory is a form of consensual non-monogamous (CNM) relationship. It is joined in this group by swinging and open relationships, but it stands as separate because, unlike the other two, polyamorous people enter a romantic as well as sexual relationship with more than one person. Amongst the wider public, polyamorous relationships are often defined by what they are not – monogamous – rather than what makes them unique and are regularly, as we have already encountered above, viewed negatively. In their 2016 paper in *Psychology and Sexuality*, a team led by psychologist Kevin Hutzler explored public perceptions of polyamory. They recruited 100 US residents, thirty-eight women and sixty-two men, and asked them a series of questions about polyamory, including what they thought it meant, what they thought about people who were polyamorous, what they thought about the practice of polyamory and whether they had ever had any interest in polyamory. The vast majority of those who responded correctly identified what polyamory was, including the vital point that it denotes multiple relationships based upon sex *and* love. When describing the characteristics of a polyamorous person, people painted a picture of someone who was promiscuous and practised unsafe sex, was lower in trustworthiness, jealousy, relationship satisfaction and morality but higher in sex drive, extroversion, physical attraction and communication skills. A third of people who responded believed they knew someone in a polyamorous relationship. Men reported higher interest in polyamory than women and were more likely to exhibit such interest if they knew someone who was polyamorous.

There is a word that floats around sometimes in poly circles called 'compersion' and that is when you feel joy and happiness at seeing your loved one happy with someone else. It gives you pleasure, it doesn't give you pain or make you jealous. **Kate**

Polyamorous people are aware of these beliefs about their chosen identity and it often places constraints on how open they are with those close to them and in the wider public arena – another example of how the public nature of love impacts the individual experiences of people. But polyamorists are keen to counter these views by defining polyamory by what it is rather than what it is not – non-monogamy – and explaining both how they feel within their relationships and the benefits it brings to them. In my discussions with people who identify as polyamorous, the concept of compersion, a term introduced by Kate above, is central to polyamorous relationships as well as the concepts of trust, negotiation, honesty and lack of jealousy.

. . . when I first came out, what surprised me most was that people were very at ease and not bothered by the idea of me sleeping with other people, but people were very bothered by me loving multiple people. That really bothered their sensibilities . . . it was shocking to

me that what people were offended by was not that I wanted to have sex with a lot of people, because I don't, but I wanted to have really intimate, emotional relationships with other people. **Jo**

In contrast to the majority belief that polyamorists place little weight on the importance of morality, people who identify as polyamorous perceive of polyamory as an ethical approach, one where the likely occurrence of relationships beyond the monogamous are acknowledged and shared openly. This is in contrast to the 'double standard' practised by a significant minority of monogamous people where illicit relationships sit alongside the socially acceptable exclusive dyad, hidden from general view. Indeed, in his paper 'Notions of Love in Polyamory', sociologist Christian Klesse argues that rather than being about promiscuity, polyamorous relationships are about commitment, about valuing long-term relationships as, in contrast to the extracurricular affairs of the monogamous relationship, every relationship is of equal standing and importance, and each member openly acknowledged and valued. Polyamorous relationships are based upon constant open communication and negotiation where rules are established about the boundaries of each relationship, and members trust each other to stick to these rules. Where rules become difficult to adhere to then they are re-negotiated, an often tricky task, to ensure everyone is still comfortable and everything remains in the open. Polyamorous people acknowledge that there are downsides to this practice – coordinating the lives of several people to ensure everyone has adequate time and investment to maintain the relationship and making sure everyone's needs are considered can be hard – but what they are clear about is that there are no limits on romantic love.

[Polyamory] is about growth and about love and the fact that love . . . I don't really think it can be contained. You can try and contain it . . . people shut themselves off from it to other people, but I definitely think you can grow from loving other people and wanting the best for them . . . you can have that with more than one person. **Rebecca**

I think the more . . . and not just romantically . . . the more you experience love and connection the more love you have to give as your relationships are fulfilling and energise you instead of exhausting you. **Jo**

As Jo and Rebecca suggest, polyamorists do not believe in 'the one', nor that love is a zero-sum game. Indeed, many conceive of their relationships as increasing manyfold the benefits that we all accrue from our romantic relationships. In her paper exploring the experiences of those in consensual non-monogamous relationships (CNMs) Director of Gender Studies at the Centenary College of Louisiana, Michele Wolkomir, interviewed ten polyamorists, six men and four women, who ranged in age from twenty-one years of age to sixty-seven. And their thoughts echo those of Rebecca and Jo. One interviewee, on being taught at high school that 'if you give everyone a piece of your heart then there is not enough left for important people in your life', thought that this perspective was 'insane . . . giving more people a piece of your heart would make it stronger and more full – not less'.

Summarising the views of six other polyamorous participants, Michele concluded that they chose to have multiple relationships because this increased the amount of connection and intimacy in their lives. They were not making up for an absence, but increasing what they felt to be the good things in their life. And the data is clear that the benefits to health, happiness and wellbeing that we see in other close human relationships extend to those in polyamorous relationships. In their review of psychological wellbeing and relationship quality in CNM relationships covering thirty-three studies, Alicia Rubel and Anthony Bogaert reported that overall those who practised CNM tended to view themselves and their lives positively and felt that CNM had improved their psychological wellbeing overall, although social attitudes and stigma could have a negative impact. As compared to monogamists, non-monogamists reported similar results on a number of psychological measures including life satisfaction, depression, personal fulfilment, obsession-compulsion, anxiety and paranoid ideation. When it came to relationship satisfaction, again there were no significant differences between monogamous and CNM relationships. When measured using the Dyadic Adjustment Scale, which includes four subscales – satisfaction, cohesion, consensus and affectional expression – the vast majority of studies showed no difference in results regardless of relationship style. Of eight studies, six reported similar results, one found that CNM had significantly higher levels of relationship happiness and one reported lower levels of satisfaction. Interestingly, in two further studies, individuals who had moved within their relationships from monogamy to CNM found that the change had improved their happiness or satisfaction in the relationship, keeping their marriage together. Alicia and Anthony concluded that there was no discernible difference in psychological wellbeing, overall relationship adjustment, jealousy, sexual satisfaction and relationship stability between monogamous and CNM relationships.

Is All Love Equal?

I don't think any parent loves their children exactly the same. They love their children but they love different things about their children and it is the same with my husband and my partner . . . I love lots of different things about them and I feel they complement each other . . . they blend together nicely . . . my husband would always get me a tissue if I were crying but my partner would open a bottle of wine . . . **Kate**

One of the public questions I get a lot still is 'OK, so you love these two people but at the end of the day who are you going to pick?'! I think people often view love as a very finite resource, specifically romance. People don't [say] 'I have multiple kids but then they don't get enough love' . . . that would be a strange thing, but romantically I think people absolutely feel that way. **Jo**

When questioned about how it is possible to truly romantically love more than one person at a time, all the polyamorists I have spoken to point to our capacity to love all our children equally at the same time. As with all aspects of love, it is hard to tap into someone's personal experience. We have to take their word for it that their love for each partner is equally powerful. But evolutionary anthropology does have a few things to say about the possibility of polyamory. It is clear from previous chapters that human monogamy is, in part, a social construct promoted to maintain the status quo, but it is also the case that where children are concerned we have evolved a system of dual parental investment which requires both parents to stick around at least until the child is of an age to fend for themselves, which arguably for the human child is well into adolescence. But this does not mean that each parent cannot have further partners outside the nuclear family. This highlights the different forms of monogamy that we need to recognise. There is sexual and emotional monogamy, but there is also social monogamy where the couple raise children together but also have other romantic relationships, although not child-bearing ones, beyond the family. The lack of childbearing in the other relationship tackles, to some extent, the issue of sharing precious resources which are argued to be the basis of the evolution of jealousy (something we will touch on to a greater extent in Chapter 9). Several of the polyamorists I interviewed for this book did feel that bringing children into a polyamorous relationship might be difficult, but this was not because it would lead to disequilibrium in the relationship or indeed be detrimental to the child's development, but because of the reaction of the outside world to their arrangement and the repercussions this might have upon the child, particularly at school.

But what does science have to say about polyamorous love? In their 2007 paper in *Hormones and Behaviour*, a team of psychologists from Simon Fraser University in Canada, led by Sari van Anders, explored the role for testosterone in monogamy and polyamory. Testosterone is the hormone that makes us competitive in the mating game; it increases our motivation to find a pair bond and, at least in men, traits which indicate high testosterone – such as a square jaw and strong body – are deemed to be more attractive. It was already known that men who were partnered exhibited lower testosterone than single men – suggesting low testosterone may have a maintenance role in pair bonding – but Sari and her team wanted to explore what the role for testosterone might be in influencing the likelihood of someone seeking a polyamorous relationship. They hypothesised that polyamorous people would exhibit higher testosterone, because the possibility of competing for a partner is always present, than those who were monogamously partnered. They recruited eleven men and thirteen women who were single, eleven men and six women who identified as monogamous, twelve men and eleven women who were polyamorous and currently had multiple partners (ranging from two to six) and six men and four women who identified as polyamorous but currently did not have multiple partners. They took saliva samples from all the participants, always at the same time of day to control for diurnal fluctuations, and set about the glorious task of analysing a lot of spit. What they found was that, as predicted, single people did have higher testosterone than monogamous people, *but* the testosterone of polyamorous people, multiple partnered or not, was even higher. The question arises: does having higher testosterone motivate you to be polyamorous, or does having more than one partner lead you to have higher testosterone? The jury is still out, as Sari's study was cross-sectional rather than longitudinal. It would take a longitudinal study, on a par with those carried out to confirm that the drop in testosterone in new fathers is due to parenthood, to unravel the direction of causation. But what was refreshing about Sari's study was that it considered people of all ethnicities and sexualities, which is unusual in the polyamory literature where the vast majority of studies are carried out on white, middle-class westerners.

The Polyamorous Brain

So we see hormonal differences between those who practise monogamy and those who practise polyamory, but this doesn't really tell us much about their experience of love; it is more about motivation. The question arises: do the results of neural studies reflect the idea that polyamorous people love their partners equally? Is there evidence of a zero-sum game or evidence that the finger-print of love is there for each and every partner? The closest we have to an answer is found in a study published in 2017 comparing the brain activity in ten men who identified as monogamous to that of ten men who identified as non-monogamous, here defined as being somebody who was or had been in a relationship with multiple partners, either as a result of cheating or consensual non-monogamy. Unfortunately for us, Lisa Hamilton and Cindy Meston did not detail the ratio of 'unfaithful' to CNM partici-pants, so we have no idea how many identified as polyamorous rather than being active swingers, living in an open relationship or conducting an affair. Their method involved showing the par-ticipants a set of four images, at random, of neutral landscapes, neutral landscapes with people, sexual and romantic images while in a fMRI scanner. For the sexual versus neutral image contrast, there were no differences between monogamous and non-monogamous participants in brain activity. However, when it came to the neutral versus romantic contrast, the differences were significant. Monogamous men showed much greater activity in the areas of the brain which are activated when we are in love – those related to reward and pair-bonding behaviours. So greater activation was seen in the thalamus, nucleus accumbens, caudate, putamen, insula and the prefrontal cortex, where the cognitive elements of love find their home. In contrast, non-monogamous men had little activation in the reward centres of the brain, sug-gesting that they did not receive the same levels of neurochemical reward as the monogamous men when viewing the images, but much more activity was evident in the cortex, implying that the romantic images required much more conscious processing for this cohort than for the monogamous group where the neural reaction was more instinctive.

So what are we saying here? That polyamorous people do not experience love in the same way monogamous people do? Well, no. In the first instance, this study, as well as being on a very small group of people, was on people who had multiple partners but these were not necessarily people who identified as polyamorous. It is a distinct possibility that some, if not all, did not experience love with multiple partners, their relationships being based upon multiple sexual but not emotional relationships. Secondly, the scanner task involved looking at romantic images rather than interactions with actual partners, so at most we are seeing a difference in the way non-monogamous and monogamous people perceive romantic love, not how they experience it. So the jury is very much still out on whether multiple, concurrent romantic love relationships are possible. There is an argument that the analogy to parental love for children or platonic love for friends, which is often presented as proof by polyamorous people, is not suitable, because having multiple children or friends does not negatively impact the chance that your genes will survive to be passed down the generations. Indeed, the more children you have, the more likely this is, and the greater your number of friends, the greater chance of a healthy, happy and long life. As a consequence, evolution has selected for us to be able to love many children and friends at once. But when it comes to having more than one romantic love there is the risk that this *does* lead to a decrease in the amount of resources your children, the carriers of your genes, receive, particularly if your partner has children elsewhere to support. This can risk the survival of your genes. As a consequence, the possibility does exist that we have evolved to focus all our romantic love on one person to maximise the chance of our children's survival (although the prevalence of affairs would suggest that evolution hasn't got its way). But at present we just do not have the empirically derived data to provide an objective answer to this question. There is much more research, particularly scanning research, that has to be done. In any event, should we dismiss the subjective reports of polyamorous people simply because they are not based upon hard, testable data? We are more than willing to take a monogamous person's word for it that they feel love without resorting to the scanner, so why do we find it difficult to accept the word of a polyamorous person that what they are feeling for all their partners *is* romantic love?

Romantic Love: The Pinnacle of Life's Achievement

At the end of the day being aromantic is just one part of who I am even though being aromantic, because it refers to romance, is seen as the most important bit of me. I have different interests and likes and dislikes and being aromantic is just one part of that. But when I am put against all of those media, TV, books, films and all those romantic things, it does feel alienating at times. **Jamie**, aromantic

It is very much the case within the west that romantic love, and being in a romantic relationship, is sold to us as the point of life. To not be cosily paired up is seen to be somehow a failure, a sadness regardless of how much you say you are happily single. This pressure is particularly there for women. Be a single forty-year-old male and you are a freewheeling bachelor, at liberty to live the life of an international playboy, but be a forty-year-old single woman and the word 'poor' will regularly be attached to your name and you risk being stereotyped as the lonely spinster, stuck on her shelf with her brood of cats. Why do we view romantic love in this way? There are so many ways a human can experience love – I hope this book is making that very clear – but we still seem to think that the most important, even desirable, is romantic. In a way this is an odd position to take. As I mentioned in Chapter 3, if you don't want to share your life with a romantic partner or partners you can easily dispense with this form of love as long as you have the powerful resource of platonic friendship love to ensure you live a happy and healthy life. Indeed, while some of us are happy not to actively seek romance, but are content to give it a try if it lands in our path, some are aware that they do not want a romantic relationship based on the knowledge that they feel no romantic attraction, and often no sexual attraction, to anyone of any gender or sexual persuasion. These are the aromantics and it is their lot to live in a world seemingly obsessed with the concept of romantic love.

The general public see aromantics as being rather cold or unloving just because they don't experience romantic love or romantic attraction. Or some people think it doesn't exist and you are making it up because being in love is about being human. People forget about the love they have for their family, that they have for their friends or even the love they have when they are eating food or the love of watching a sunset or stargazing, and those things can be called love, but in the moment romantic love is the only proper love. **Jamie**, aromantic

Aromantics are the group of people who do not feel romantic attraction to another human. In many cases they are also asexual, meaning that there is no sexual attraction either. They are a group of people who have remained pretty well totally hidden until very recently, when it became apparent to the press that support groups of aromantics littered online social-media platforms such as Reddit and Tumblr. This has led to a small number of articles in the press, but the academic literature is largely silent on this particular identity. What is quite clear from my conversations with people who identify as aromantic is that, contrary to popular belief, aromantics are not unloving, they experience love for their friends, their family, their pets just like the majority of people. Nor do they want to live lonely lives. Many want to cohabit and have children, but they want the basis of the relationships that underpin these to be platonic – this is known as having a queer (as in different from the norm) platonic partner (QPP) – and online forums catering to aromantics often have areas where aromantics can search for a suitable QPP.

Yeah, I definitely know I want a partner, it's just not going to be a romantic partner. So finding a QPP is a goal for me. I want . . . I don't want to live alone, I want someone to live with . . . just someone you can have that lifelong connection with and you are partners . . . but it is not in a romantic sense or a sexual sense. **Sara**

But in a world whose culture – media, art, books, music – is obsessed with romantic love – reflecting it, questioning it, mourning its loss – it can be hard for aromantics to find a social group which understands them. Even within the queer community, which is arguably better at embracing difference than mainstream society, people struggle to understand someone who has no sense of romantic attraction:

I went to Pride and there was all this thing about love, love is great, everyone has love, but it felt a bit like I was being excluded. It wasn't intentional but I don't feel the love you are talking about. So even then it was a bit uncomfortable to think 'Oh well, these people are all connected on the fact that they are going to experience romantic love in any form and I am on the outside because I am not.' **Sara**

And closer to home it can be hard to come out to family and friends because of a lack of acceptance and understanding:

I don't tell my extended family, I don't tell my colleagues, because I know the chances are they are not going to understand or accept it. And it's fine. I can field the questions of 'Ooh, when are you going to start dating someone?' [When I told my mum] at first she thought something had happened that had resulted in this and then she was worried I was going to be alone. [My parents] are happy in their relationship so they want me to be happy in my relationship even though they don't understand that that is never going to be something I will be happy with. You get it a lot: 'You just haven't found the right person.' **Sara**

In terms of my family, I think they are kind of disappointed actually when we talk about it. They had hopes for me that I would get married, but as I don't feel romantic attraction it's going to be pretty hard as that is how most people do it. To be fair, I think they think it would make me happy having a partner. **Jamie**

But to aromantics the explanation is very straightforward, and while the idea of romantic love to them is at the very least uncomfortable – 'if I think of myself in [a romantic relationship] then it is a case of "Oh no. No thanks." I feel when I think of it ... it doesn't feel like me ... [and] I do feel repulsed thinking about being in a romantic relationship. I wouldn't be being me if I were in a romantic relationship' (Jamie) – they have no issue with those around them finding happiness this way. In the same way as polyamory or monogamy might constitute part, but not all, of a person's identity so aromanticism is part of Sara and Jamie's identities. But as they are keen to stress, it is only one part and there is much more to say that is not related to their relationship status. However, while we continue to privilege romantic love above all others, aromantics will remain a curiosity and be the target for many, at times insensitive, questions as we struggle to understand, in a world seemingly obsessed with romantic love, how anyone can live a happy and contented life without it.

What links polyamory and aromanticism is our enduring fascination with both, because they subvert our deep-seated ideas about romantic love. They introduce the possibility – which is very hard for some to take – that there is no 'one'. Indeed, there might be 'the many' or no one at all. The stories we have been told since we were children, the pop songs which unleashed our tempestuous emotions as teenagers and the films which we cry over as the protagonist does or does not achieve the promised land of coupledom may not be an accurate depiction of reality. This destabilises many, but for some who struggle to fit themselves in the box that has been dictated by society, realising that there is a whole spectrum of possibility out there is wonderfully freeing.

Exclusivity, or not, has been the subject of this chapter. We've explored the experiences of those who identify as monogamous, polyamorous and aromantic, had a good look at their physiology and their brains and considered how they are perceived by wider society. However, the next chapter deals with perhaps the most private love of all, that between ourselves and our god or gods. Close relationships are made closer still by our willingness to be emotionally vulnerable in front of our love; revealing our darkest thoughts, our deepest worries, our greatest fantasies or dreams. And maybe we are free to do this to its greatest extent when we commune with, when we offer our love and thoughts to, a sacred other.

CHAPTER EIGHT

SACRED

● ● ● ●

'Thou shalt have no other gods before me'
The Ten Commandments

*'No other love may override one's love for God; God should be
the highest and foremost object of love.'*
Islam

*I love God now as a woman loves. I love and I know I am
loved and beloved.*
Niamh, Catholic nun

In 2010, 5.8 billion adults and children were religiously affiliated;
84 per cent of the total global population. According to a survey of
more than 2500 censuses carried out by the Pew Research Center,
2.2 billion people identify as Christian, 1.6 billion as Muslim,
1 billion as Hindu, 25 million as Sikh and 14 million as Jewish.
Even where no religious affiliation is identified, many people
report having a spiritual life and believing in a god or gods.
And for many the teaching of their faith makes it clear that the
first and most important love is the love one has for God. This

perhaps reaches its acme in the exclusive love that exists between the Christian God and His most dedicated monastic followers who abstain from many of life's normal trappings, and potential sources of joy, to give their lives in their totality to Him.

In this chapter I want to explore what we know about religious love. If we turn to the hard sciences – neuroscience and genetics – the answer is very little, although scanning studies are starting to explore what happens in our brain when we commune with our god. But our knowledge of what human relationships entail – trust, empathy, attachment, reciprocity, maintenance – enables us to question whether these are characteristics that we can identify in the relationships that exist between individuals and their god. Both published studies exploring religious attachment and my own conversations with those who have chosen a religious vocation feed into this. For me, the interesting question is whether what these individuals feel for a god is akin to the love we have for our fellow humans. Does the neural fingerprint of love light up the scanner screen? And if these are relationships of equal weight to those we may have with our friends, lovers and families, do they bestow the same benefits for health and life satisfaction which are the real value of having love in our lives?

God Is Love

All world religions have something to say about what love is, which has shaped both our individual perceptions of love and, as we know from Chapter 6, wider cultural attitudes to it.

The monotheistic religions share an understanding that love of God is the most important love – beyond family, beyond spouse, beyond neighbour – but the nature of the expression of that love differs.

'So faith, hope, love abide, these three: but the greatest of these is love.'

1 Corinthians 13:13.

Within the Catholic church, love is coupled with charity and is seen as the greatest theological virtue, and within the broader Christian church the relationship one has to God is a highly personal inter-human interaction. In Judaism, love is equated with kindness and helping those in need regardless of their identity, Jew or non-Jew. Love for God is represented in faithful adherence to the ways of life laid down in the Torah. In Islam, love comes in divine and human form but should be understood from a position of truth and enlightenment; our love for another should not cloud our judgement. In Hinduism, love is *Kama* and is associated both with devotion and pleasure, while in Sikhism love is based in marriage and the family.

I think that Jewish understanding about love for God is fundamentally a different kind of understanding from a Christian one. Judaism is very strong on teaching that one's relationship to God is not primarily a private experience, which is why Judaism is closer to Islam . . . if you talk to Jewish people . . . religious people, you don't tend to hear people talking about their love of God in the way that you would if you talk to Christians. I don't have a belief in personality relating to the divine. So it is quite hard then to narrow it down to have that sense of 'love of' . . . it is impersonal.
Benjamin, rabbi

However, differences exist between the three main monotheistic religions in the possibility of a personal, exclusive relationship between the individual devotee and God. The possibility of a love as we conceive of it in close, dyadic human relationships. Whereas the relationship between Christian followers and God is highly personal and possible without intermediaries, Jewish and Islamic theologians argue that the relationship with God in Judaism and Islam is achieved through intermediaries such as the rabbi or imam. So congregation with God, rather than being found through the Christian practice of personal prayer which is akin to an un-ritualised and free-flowing conversation, is found through ritual and the carrying out of good deeds within the community setting. Further, neither religion conceives of God in the anthropomorphic terms of Christianity. He is not viewed as being an individual 'personality'; a friend, father or lover. And because only Christianity considers a personal relationship with God to be an integral factor in devotion, I will rely mostly on evidence from Christians to consider whether love of God is the real deal.

God As Attachment

At the basis of all of the most powerful human loves is attachment, the profound and intense bond which is rare but, when secure, enables you to take on the world. One of the pioneers of attachment theory, Mary Ainsworth, listed four criteria which mark out a relationship as an attachment: a desire to maintain proximity, viewing the attachment as a secure base from which to explore, perceiving the attachment figure as a safe haven and experiencing separation anxiety when parted from the attachment figure. I argued earlier that for me attachment is one of the key markers of deep love, signalling the intense and enduring companionate love that will see us through to the end of our lives in contrast to the passionate sensations – sexual and platonic – which occupy the earliest stages of a relationship. We experience attachment with our parents, with our families, with our children, with our close friends and even with some of our pets, but can we experience it with God?

In the early 2000s psychologists Richard Beck and Angie Mc-Donald set out to explore two questions. First, was the relationship that Christian devotees had to God an attachment? Secondly, could they develop an assessment tool – akin to the Experiences in Close Relationships (ECR) measure that we met in Chapter 2 for romantic relationships – to measure this? In the first instance, this meant recognising the potential for the two dimensions of attachment – anxiety of abandonment and avoidance of intimacy – in the relationships that devotees had with God. For Richard and Angie's study, these devotees numbered 507 students from Abilene Christian University, 62 per cent of whom were female and 85 per cent of whom were Caucasian. Beck and McDonald developed a seventy-item trial Attachment to God Inventory (AGI) which included questions such as 'My experiences with God are very intimate and emotional' (intimacy) and 'I often worry about whether God is pleased with me' (anxiety).

Having asked their 507 religious guinea pigs to complete the trial seventy-item measure, Richard and Angie carried out some analysis. In the first instance, it was clear that the relationships that had been reported to them could be conceived as attachment relationships; the four crucial dimensions laid out by Ainsworth were present. Secondly, having carried out a principle components analysis, to explore the most robust and dominant factors in religious attachment to God and to ensure that the two dimensions carried equal weight, they produced a much more manageable twenty-eight-item AGI comprising fourteen questions relating to avoidance and fourteen to anxiety. Next, they went on to test *this* twenty-eight-question AGI on a second group – 118 students this time, eighty-nine female and 72 per cent Caucasian – to ensure that their results from their first study were replicable – that religious attachment was a real phenomenon – and to compare them to how their participants scored on the thirty-six-item ECR for romantic relationships. This second area of interest was included to try and answer one of the long-held questions in the psychology of religion – does an attachment to a god act as a compensation for a lack of attachment to human figures (compensatory) or does it mirror the nature of the human attachments in the person's life (correspondence)?

Their results showed, in the first instance, that the AGI was a robust measure for attachment to God (results were replicated from Study 1). When considering whether the nature of the attachment was compensation or correspondence, the results of this study pointed to correspondence on the anxiety dimensions – so a secure ECR predicted a secure AGI, an insecure ECR an insecure AGI – but there was no clear trend when it came to avoidance. The relationship was weakly positive – so again, low avoidance in the ECR mirrored low avoidance in the AGI – but not significant. However, these results would suggest that a love for God is not motivated by the need to compensate for a lack of love in one's earthly life but by a genuine attraction to God as a valuable member of a social network.

Finally, Beck and McDonald embarked on their third and final study. Here they widened their focus to include the local Christian community, recruiting 109 members from church-led adult-education programmes around Abilene in Texas. These were thirty-eight members of a Church of Christ congregation, thirty-four members of a Catholic church and thirty-four members of a non-denominational, charismatic congregation. 61 per cent of the participants were women, with 79.8 per cent being Caucasian, 11 per cent Hispanic, 2.8 per cent African American and 2.8 per cent Asian American.

This group was asked to complete the twenty-eight-item AGI, the thirty-six-item ECR, a second measurement of romantic attachment (the Relationship Questionnaire, included for robusticity) and the twenty-item Spiritual Wellbeing Scale. Here's hoping there was a lot of coffee and free biscuits on offer to see them through. Their results showed yet again that the AGI was a robust measure of attachment to God, but what was really interesting, and is played out in my interviews with religious people across the lifespan, is that the older community population were more likely to be secure in both their attachment to God and their romantic attachment as compared to the students, who were much more likely to be more anxious and more avoidant in both their religious and romantic attachments. As in our other life relationships, it would appear that the early years of a relationship with God can be a time of turmoil and anxiety as you find your feet and become secure in your relationship with Him.

We Are Family . . .

I think of him now as Father and I think of Jesus as Brother . . . I was listening to an American Jesuit talking and he was talking about the disciple Jesus loved and he made very clear it was never identified in the scriptures . . . So I am the one Jesus loved. I was able to nuzzle up to Him at the Last Supper, and also be there at the cross and then be given Mary as my mother and that means I am Jesus's little sister and I think of myself as that ... that is currently what motivates how I am. **Niamh**, Catholic nun

In the Christian tradition, God is referred to regularly as Father, Mother and even lover. Catholic nuns famously become 'Brides of Christ' and wear a wedding ring. My interviews with those who have a religious vocation within the Christian church reflect this. For some, he is a loyal and much-relied-upon friend, someone to whom you can talk about anything, safe in the knowledge that you will still be loved. For others, like Niamh, he is the head of a family, with a mother and a brother to boot. But in all cases what is envisioned is a personal dyadic relationship which can sit comfortably at the centre of a follower's social network alongside their more tangible, survival-critical relationships.

... by now we go back a long way together and I can be grumbling to God when life is very boring and normal. It's like having a friend who you don't have to call up on Zoom, you know? Because he is always there. When everyone's around the house, you are not particularly registering them but they are there. And with God he is always there. **Sister Agnes**, Church of England nun

But the question remains – is love of God the same as love for family or friends? My interviewees report that the nature of their attachment to God, while comparable, is different from that found within the earthly realm. For many, it is the source of absolute security and openness. God can be told anything and he will never reject you. And while you may fall out with family, friends and lovers, there is no risk of this happening with God. His love is the ultimate secure base. Here are Amy, Niamh and Sister Agnes on the subject of the difference between love for God and love for our fellow humans:

He is the one who will help you and be with you all the time and I find it is deeper than any human relationships I have had. You learn that He is just as important as a human being . . . He is still your friend and He needs the time you would give to someone else and the enthusiasm you give to other people . . . it is probably more rewarding than a human relationship because of all the good things you receive and the blessings, everything is more abundant. **Amy**, Catholic nun

. . . there are certain things I would never tell even my dearest friends . . . but I would talk about it with God. So God knows the bitchiness and the human frailty that is me and the sinfulness of me, [He] loves the worst as well as the best of me. God knows all the impulses and the dreams and the hopes and the unfulfilled stuff. He knows all of that. **Niamh**, Catholic nun

The huge difference is that love for God is less precarious. I was very lucky to have parents who loved me . . . while they were alive they were anchorage for me . . . I haven't had a home in that sense since my father died. With my children, there's my own inadequacies and there are misunderstandings and you are left wondering if it matters and if the relationship could end. Never from my end because I would always love my children . . . to be a parent is to run the risk that your children will shed you. To be romantically connected is to run a very high chance that the person you are romantically linked to will shed you and not be interested any more . . . with God, he is always there. **Sister Agnes**, Church of England nun

Now the argument could be put forward that for a nun, who has committed her life to God, comparing love of God to some human loves, such as romantic or parental love, is not within their skill set, so how can they draw these conclusions? But let me tell you about Sister Agnes. When I first contacted a number of convents looking for nuns who would be willing to speak to me for this book, she was one of the first to get in touch. But she was doubtful about her suitability because she had come to the convent later in life after living a life which included a husband, children and grandchildren. I went back to her and said, on the contrary, she was gold dust! Here was someone who had the life experience to compare love of God with the types of love which many of us experience in our daily lives. She was my opportunity to try and get even a small handle on whether we can place religious love on the same spectrum, in the same box, as that for a human. So for me her answer to my question about comparing love of God to the other loves in her life, quoted above, is hugely valuable. Beyond this, perhaps her answer to my question 'what is the most powerful love you have experienced in your life?' is the most telling when it comes to the nature of her love for God. This is a woman who has had so much love in her life but her answer was a single phrase:

From and for God.

From the Christian perspective, love of God is an attachment with all the characteristics of more earthly bound attachments. Proximity is found in personal prayer when followers have a private conversation with God, free of the rituals of collective prayer and worship – 'I did a retreat on a Sunday so every hour you stopped to pay attention to God, to pray to God . . . and that day was way better than any other day because you were consciously stopping . . . it is about forming a passion . . .' (Amy). Secure haven and a secure base are sought when life gets tough – 'you could just be yourself [with God] and know he loves you as you are. You don't have to perform. You don't have to be a good girl or a good nun or whatever . . . it made you a better person. It made you feel better' (Niamh). And separation distress ameliorated due to the low levels of anxiety about abandonment that a close relationship with God brings to his followers – 'I feel heard [by God] . . . one of the things that I have discovered is that I don't actually know what it would be like to be completely on my own . . . I have never felt forsaken by Heaven. I've never felt He left me on my own' (Sister Agnes).

It would appear to be the case that the love between God and his Christian followers, at least where those with a vocation are concerned, is an attachment and an attachment which stands up well when compared to those we have with lovers, friends and family. But to this point the data we have explored is all subjective; we have to take the results of the interviews and questionnaires at face value. To truly place religious love in the same category as other human loves, we need to bolster the argument with further evidence, ideally objective evidence. So let's welcome to the table the neuroscience of religion.

How Do You Fit Fifteen Nuns in a Scanner?

In early 2000, Canadian psychologists and neuroscientists Mario Beauregard and Vincent Paquette set about exploring the neural correlates of a Christian 'mystical experience' – a form of intense personal prayer. They recruited a group of fifteen Carmelite nuns who were willing to attempt to recall an instance of union with God while in a fMRI scanner, a not insubstantial task. The average age of the group was around fifty and all had been in the Order for between two and thirty-seven years. Each nun was asked to lie in the scanner, eyes closed, and relive the most mystical experience they had had while being a Carmelite nun. This was the task, rather than asking the participants to form a union with God there and then, because the nuns made it clear that, bearing in mind it was a reciprocal relationship and God might have other ideas, they could not summon Him at will. Alongside the 'mystical' condition, participants were asked to take part in a control condition – lying with their eyes closed in the scanner and recalling the most intense state of union they had felt with a fellow human while in the Order.

Following scanning, the nuns were asked to score the intensity of their union on a zero to five scale where zero was no experience of union and five the most intense union ever experienced. In addition, they rated their experience on the Mysticism Scale, developed to capture the phenomenology of the encounter, and undertook an interview to explore their subjective experience. The average intensity of the mystical condition was 3.06 and that for the control 3.04, pretty much on a par. The most significant comments selected from the Mysticism Scale were 'I have had an experience in which something greater than myself seemed to absorb me', 'I have experienced profound joy' and, most powerfully, 'I have had an experience which I knew to be sacred'. And in interview, several nuns mentioned that during the mystical condition they felt the presence of God; a feeling of His unconditional and infinite love and a sense of plenitude and peace.

When viewing the scanning results, it was clear that mystical union employed three distinct areas of the brain; those related to visual perception, cognition and emotion. But what was most striking was the areas of the limbic system and cortex which showed peak activation. In the cortex these were the medial prefrontal cortex (MPFC), the anterior cingulate cortex (ACC) and the orbito-frontal cortex (OFC) – all key areas within the prefrontal cortex which are recruited during social cognition. Within the limbic system the caudate nucleus, which has crucial links to the prefrontal cortex, lit up. Does this pattern ring any bells? These are the activations of love which we first encountered in Chapter 2. The neurochemical reward of the unconscious brain which motivates us to, and rewards us for, maintaining our survival-critical relationships, the conscious contemplation of the cortex which allows us to reflect upon our experiences and attempt to encapsulate them in words and the critical neural link between the two enabling us to have a truly human experience of love. It would appear that when Carmelite nuns interact with their God, what they feel *is* love.

God As Human

But is this love really a parallel for the relationships we have with our fellow humans? Does the characterisation of God as a friend, father or lover mean that when they commune they really feel they are interacting with another human being, albeit one not in physical form? In 2009 a team of theologians, neuroscientists and anthropologists – what a great mix – set out to explore what happened in an individual's brain when they talked to God. This time, the cohort was not made up of those with a vocation but the general congregation of a Danish Lutheran Church. Twenty healthy young male (six) and female (fourteen) volunteers were recruited and asked to carry out four distinct verbal tasks in the scanner lasting thirty seconds each. The first two were recitations of the Lord's Prayer (religious formalised speech act) and a nursery rhyme of their choice (secular formalised speech act) while the last two were a period of personal prayer (religious improvised speech act) and a wish list to Santa Claus (secular improvised speech act). The tasks were presented in a random order. The aim of the contrasting religious tasks was to understand the potential differences between undertaking institutionalised, ritualised worship and non-institutionalised, personal worship. The nursery rhyme and list of Christmas wishes acted as controls to ensure that any unique activations viewed in the religious conditions were as a result of their reference to God rather than simply a product of the nature, improvised or formalised, of the recitation. It should be made clear that all the participants definitely believed in God, and believed their relationship to God was reciprocal (measured by their level of agreement to the phrase 'I am absolutely sure God reacts to my prayers'), but had realised (spoiler alert; cover the eyes of any children) that Santa Claus does not exist.

What did they see? For the two formalised speech acts – the Lord's Prayer and the nursery rhyme – areas of the brain linked to rehearsal and retrieval were activated. But the most striking result occurred when comparing personal prayer to both wishes to Santa Claus and the Lord's Prayer. For both of these comparisons, what jumped off the scanner screen was that for personal prayer, but not the other two, areas in the medial prefrontal cortex (MPFC), the left temporal-parietal junction (TPJ), the left temporopolar region (TPR) and the left precuneus were active. Why is this significant? Because the MPFC, TPR and TPJ are not only associated with social cognition but are fundamental to mentalising – remember, that is the ability to infer someone else's mental state – and reciprocity. The study's authors, Uffe Schjoedt, Andreas Roepstorff and Armin Geertz, conclude that these results show that talking to God, who the participants believe is real, rather than the fictitious figure of Santa Claus, is akin to talking to a real person. Their argument is strengthened by the fact that previous studies asking participants to play online games against either a human or a computer recruit their TPJ, MPFC and TPR when interacting with the human but *not* the computer. It would appear that Christians view God as a person with agency, his own thoughts, motivations and desires and that their relationship is as two-way as any they experience with their fellow humans.

Relationship Investment: It's All About Time

It would appear that the relationship between Christians and God is an attachment, has the neural hallmarks of love and re-cruits the same areas of the brain as when we take part in social interaction with our friends, family and lovers. And the young Christians who made up the cohort in Uffe's study made it clear that not only was their love for God reciprocated but that personal prayer – which I think we can describe as a conversation with God – was an important way of maintaining their relationship. On average, they involved themselves in personal prayer twenty times a week. And as we know from previous chapters, reciprocity, and other forms of maintenance, are key to ensuring that a relationship endures. Just try only receiving in a relationship or failing to give it adequate time and you will find that it comes to a pretty rapid conclusion. As a consequence, one of the key questions I asked those with a religious vocation was what they gave to God and He to them. Here are a few answers:

I let God love me. Maybe that's the greatest thing I could give him because He is all gift and He gives and gives and gives and gives . . . when He was creating the world He thought of me, so I am from love, of love, for love. I give him my desire always to have Him as the first and most important . . . There is an Ignatian prayer . . . 'I am thine for all eternity' . . . **Niamh**, Catholic nun

I experience God-given peace of mind. I experience peace and tran-quillity. What do I do for God? It is very hard to know when you are doing God's will and I sometimes think maybe I'll get a shock when I die and find I wasn't doing things the right way at all . . . I try to give time to God . . . I see God very much in people and the people that we serve. **Saoirse**, Catholic nun

We always say we are the body of Christ, we are the face of Christ, so when I do my ministry I try to do His work . . . be like Christ to people . . . I receive so many things. It is crazy what I receive . . . support from other people who I wouldn't even think of . . . He introduces them into my life . . . for me, the list is endless. **Amy**, Catholic nun

This reciprocity is so integral to the relationship with God that in the Catholic church nuns undertake a daily 'Examine' at the close of the day to explore when they were present to listen to God and respond to Him and where they may have missed Him that day, something our more earthly relationships might benefit from now and again:

For fifteen minutes . . . you sit and you thank God for the day and you ask Him to show you where you have met Him today and been

grateful for that . . . where did I meet You and respond? Where did I not meet You? It is about God was offering me His presence and did I accept or ignore it? You say to Him 'I am sorry about that. I know You are not going to withdraw from me, so help me tomorrow.' **Niamh**, Catholic nun

All the nuns I spoke to understand the critical role for time in maintaining a relationship. We know from Chapter 1 that one of the two constraints on the size of our social network – the other being cognitive capacity – is the availability of time, and this is because we have to commit time to nurturing our relationships to ensure they remain strong and healthy. For religious devotees, time, and time in personal prayer in particular, is key to maintaining their relationship with God and signalling how critical He is to their lives:

I think the nature of it is the more time I make to spend consciously engaged with God, the more benefit I get . . . Just to spend time with no other responsibility than to sit and be there with God . . . it's like being a half-finished statue sitting here where the sculptor puts you and waiting for the sculptor to come and do the next bit. **Sister Agnes**, Church of England nun

[It is about] *turning up and being there and if you said you were going to do something then faithfully doing it. It is really an expression of your love, a desire to grow in love.* **Niamh**, Catholic nun.

God-given Good Health

We learned in Chapter 1 that love is so integral to our survival as humans that the phenomenon of biobehavioural synchrony has evolved to ensure every mechanism in our body – behavioural, physiological, neurological – is recruited to ensure the success of our closest relationships. Unfortunately, I am not about to report on a cutting-edge study that has enabled us to manifest God in physical form and go after him with blood-pressure monitor, sample tube and scanner to confirm whether or not a relationship with God carries this important hallmark of love. But the proxy for this phenomenon can be seen in part in the link between the nature of our closest relationships and our health. We know, again from Chapter 1, that there is robust data to show that those who have close, healthy relationships in their life experience a reduced risk of mental and physical ill health, live longer lives, have greater life satisfaction and recover more quickly following bouts of illness. Can the same be said for those who reserve a place in their social network for God?

Psychologists have pondered on the impact of religious belief on psychological wellbeing for over a century. Some, such as the psychotherapist Albert Ellis, have concluded that adherence to religious belief is institutionalised irrationality and has a deleterious impact on mental health. While others, such as Carl Jung, believed that religion has a role in providing stability and meaning in an uncertain world, leading to positive mental-health outcomes. However, only in the last couple of decades has this relationship been the subject of empirical study. One of the first meta-analyses of this phenomenon was carried out by Charles Hackney and Glenn Sanders in the early 2000s. Until this time, the small number of studies that had been carried out had led to widely differing conclusions; religious belief had a positive, negative and inconsequential impact on mental health. Hackney and Sanders put this discrepancy down to the wildly different measures of both religiosity and mental ill health employed by researchers, which made comparing outcomes nigh on impossible. In their 2003 study they selected 35 studies that had explored the relationship between religiosity and everyday psychological adjustment. They then set about coding the studies to produce a list of common religious variables resulting in three clear and consistent categories of religiosity: institutional religion (participation in church activities, religious services and ritual prayer), ideological religion (ideology, belief salience and fundamentalism) and personal devotion (emotional attachment to God, personal prayer, devotional intensity). Mental-health outcomes were also coded into three categories: psychological distress (depression, anxiety, etc.), life satisfaction (self-esteem, happiness, etc.) and self-actualisation (identity integration, existential wellbeing, etc.). Once these categories were firmly and clearly established, analysis could begin.

Their results showed that overall religiosity had a positive impact upon psychological wellbeing, meaning that the chance of suffering from mental ill health is reduced if you are religious. When considering the three categories of religiosity, the weakest relationship was between mental good health and institutional religion, while the strongest was with personal devotion. Ideological religion sat between these two. For our exploration of religious love, this is a powerful finding because it shows that it is the *personal* relationship with God, the attachment that exists between Him and his devotee, which has the greatest impact on psychological wellbeing rather than, for example, the social capital that membership of a congregation might bestow. This is akin to the impact that our close human relationships have on our mental and physical health. For the believer to benefit significantly from their relationship with God, they must engage in more than the ritual of church services and prayer, and work to establish an exclusive, personal relationship with God the individual.

This idea that one of the key factors in the link between good health and religiosity is the personal relationship with God – the love for and from God – is confirmed in a cross-cultural study carried out by a team of clinical psychologists from Columbia University in New York. Led by Clayton McClintock, the team collected data from 5512 participants, 41 per cent women, who were residents of China, India and the United States. 20 per cent of the participants were Buddhist, 21 per cent Christian, 11 per cent Hindu, 2 per cent Muslim 26 per cent non-religious and 9 per cent other. The aim of their study was to answer two key questions. Firstly, what are the basic characteristics of spirituality which are significant in all three populations? They focused on spirituality rather than religion to ensure they captured the experiences of those who have spiritual belief but do not involve themselves with an organised religion. Secondly, they asked whether any specific spiritual dimensions had a significant role to play in mental wellbeing?

To ensure that they were able to capture any cross-cultural variation in spirituality, Clayton and his team scoured the literature and brought together over 150 different self-report measures of spirituality. They carried out analysis to identify the key factors within these measures which resulted in the administration of fourteen questionnaires to their, I really hope, eager cohort. This group were also given three self-report mental wellbeing questionnaires, including the PHQ-9 for depression and the GAD-7 for anxiety. Once they had their pool of data, they carried out an exploratory factor analysis and embarked on exploratory structural equation modelling to identify the key universal characteristics of spirituality. These were religious and spiritual reflection and commitment (including commitment to a community and practices), contemplative practice (such as meditation, yoga and contemplative prayer), unifying interconnectedness (the conscious connection between people and other forms of life), love (love for self and others, including sacred love) and altruism (the helping of others). For all five factors there was a significant relationship to the risk of developing a psychopathology including major depression, suicidal ideation, alcohol and drug addiction and generalised anxiety disorder. For all three countries, this relationship was negative when focusing on the categories of love, interconnectedness and altruism. So feeling love for self, others and God, being connected to the world beyond oneself and helping others all decreased the chance of experiencing a psychopathology. When it came to the other two categories, commitment and contemplative practice, this inverse relationship held true for the populations from India and the US but not China.

It is clear from the above study that it is the social aspects of spirituality, including one's personal attachment to one's god, which have the greatest impact on the risk of experiencing poor mental health. Some of this will be down to the social capital that being a member of a church affords – a social network which is there to provide support, help and a sense of connectedness – but it is also clear from McClintock's study, and that carried out over a decade earlier by Hackney and Sanders, that it is making a space in that social network for a god or gods with whom you have a close personal relationship, akin to that with a fellow human, that is also the key. Indeed, in McClintock's study those who experienced spirituality as love for self, others and god(s) had 9–60 per cent less likelihood of having a major depressive disorder, 35–49 per cent decreased likelihood of suicidal ideation and 23–62 per cent decreased likelihood of generalised anxiety disorder. And where there appear to be exceptions to this rule, as is the case with China in the cross-cultural study, it is likely that the negative effects of spirituality on mental health in this cohort – limited to the spiritual dimensions of commitment and contemplative practice – are more likely to be associated with the less than welcome environment for religion in China, meaning that practising one's faith can be a source of considerable stress and fear.

The Cult of Celebrity

As an anthropologist whose aim it is to understand human experience, I am not particularly engaged in the debates about whether or not God exists. What I do know is that religion has been an element of human experience for at least 8000 years; the first archaeological evidence for organised worship has been found at the temple site of Gobekli Tepe in Anatolia, which dates to between 8000–10,000 BCE. Here is not the place to debate religion's evolutionary origins, though suffice to say it in all probability emerged as a way to understand the world, and subsequently as a form of social control. It can be argued that religion and, in particular, religious difference has been used to justify the most horrific acts of terrorism and war, but for many it is a source of great support and comfort and provides a set of rules and values by which to live life and understand the wider world, which ultimately reduces stress and enhances lives. However, if you do not believe in a higher being, then it can be hard to understand how love for God is any different from love for any other imaginary or untouchable being; a character in a book, a pop or film star or even an avatar in a second-life game. We have all, I am sure, experienced a connection to someone we have never, and probably will never, meet. These relationships are known as parasocial relationships; one-way connections, unreciprocated and unacknowledged, characterised by the instigator's commitment of time, energy and emotion alone. And in a world of mass 24/7 social media, some would argue that celebrities are our new gods and their worship our new religion.

Parasocial Love

Being a Stan makes me feel very happy. It makes me feel like I have something special . . . You have a treasure that no one else has. It is so easy to relate to them and be with them. It is like a friendship. These girls will always be there with me. They are something significant in my life which I really wouldn't like to lose **Harmony**, Little Mix Stan

I can still remember the moment I heard that Robbie Williams was leaving Take That (I know I am dating myself here). I was on a ferry returning from a family holiday to France in July 1995 and I knew that his departure was the beginning of the end for the band. Likewise, I can remember the moment I heard that the actor River Phoenix, my generation's James Dean, had died of a drugs overdose outside the Viper Room in LA. I was sitting in my teenage bedroom one evening when the news flash came on the TV. In contrast to the departure of Williams, which was at most a slight sadness for the end of an era and the loss of a good voice, the pain I felt in my heart for the loss of Phoenix, who seemed to sum up the angst of teenagehood and be a fellow traveller on life's journey, was real and lasted weeks. I can still summon up the sensations of the moment today. How could this beautiful, talented and apparently troubled soul be taken from us? What I felt for

Phoenix would probably have been dismissed then as a teenage crush, but now, decades later, I know that what I felt for him was a parasocial relationship with, some would argue, the potential for the same feelings and benefits of any friendship or romantic relationship.

Professor Gayle Stever has studied parasocial relationships for three decades. During this time she has spent many hours 'in the field' meeting with fans, interviewing them and analysing their letters to their idols, attending fan events and collating their extensive memorabilia collections. What she has come to conclude is that there are three levels of parasocial relationship which mirror the spectrum of relationships we can have in the real world. So the least intense, comparable to an acquaintance, is a parasocial interaction. These occur while we interact with the celebrity; reading an article about them, watching a film or contacting them via social media. We have an emotional investment in the interaction but this ceases once we exit Snapchat or turn off the TV. I think this describes me and most of the celebrities on Instagram; entertaining while they are there, and excellent fodder for my nosiness, but I can leave it at that. The second level is a parasocial relationship. I think this is where my relationship with Phoenix finds its natural home. Here you continue to ruminate on the celebrity even when you are not actively reading about them or interacting with them on social media. My grief for the loss of Phoenix continued even after I had consumed everything I could find to read about him. The third level, and this is where love might come in, is a parasocial attachment. In this case, the relationship bears all the hallmarks of an attachment. You crave proximity to the celebrity, you view them as a safe haven and source of support and you experience distress if you are unable to read about them or contact them.

The Life Story of a Crush

I could relate to the girls because I have been bullied. I just felt I had a connection to them and it was something special. I listened to some of their songs about women's empowerment and loving your body and it was such a powerful concept. I just realised this is actually a safe place for me to be myself, be accepted, I won't be judged for who I am. **Harmony**, Little Mix Stan

Parasocial relationships have their place at all stages of our life course. When we are pre-schoolers, attachment to cartoon characters is the norm and these can help us to start to explore relationships, our developing identity and prosocial behaviours. When we reach teenagehood – the time when we are most likely to accuse someone of having a 'crush' – they serve a vital role in helping teenagers to explore love and sexuality safely, to identify desired traits in future lovers and to find much-needed developmental support which may not be available in real life. Adolescence is a crucial stage of life when our focus of attachment shifts from our parents to our peers. And with American teenagers spending on average nine hours a day engaged with media of some form or another, there is considerable opportunity for one or more of these attachments to be developed with a celebrity. Indeed, where teenagers are yet to identify their peer group, or where they struggle to find peers with whom they can explore particular aspects of their development, parasocial relationships have an important role to play in their social network. In his 2018 paper, Bradley Bond from the University of San Diego explored the role for parasocial relationships in the lives of straight, lesbian, gay and bisexual adolescents. He recruited 321 heterosexual (74 per cent female) and 106 lesbian, gay and bisexual (60 per cent female) middle- and high-school students. Each student was asked to identify their favourite media personality and then answer a range of questionnaires relating to them, including the strength of their parasocial relationship with them, how often they were exposed to them via different media outlets, how similar they were to them, how attracted they were to them and how likely they were to turn to the media person, if they could, for guidance, advice or information as compared to approaching other people such as parents, best friends or teachers.

Results showed that regardless of sexuality favourite media personalities were most likely to be TV or film stars and across all groups the actress Jennifer Lawrence came out as the top pick. However, after this, their preferences diverged. For heterosexual teens, Taylor Swift, Harry Styles, Sheldon Cooper from *The Big Bang Theory* and Miley Cyrus made up their top five. For LGB teens, the diver Tom Daley, The Doctor from *Doctor Who*, *Queer as Folk's* Randy Harrison and *Supernatural's* Dean Winchester made up the top five. LGB teens were much more likely than heterosexual students to pick a media celebrity who was LGB regardless of the celebrity's gender, and the likelihood of establishing a parasocial relationship with an LGB celebrity increased if the student lacked access to real-life LGB peers, meaning that the parasocial relationship was serving a vital social and developmental need. Further, LGB teens were seen to see their chosen media celebrity as a reliable source of support, unlike heterosexual adolescents. Bradley concluded that LGB teens rely on their parasocial relationships with LGB celebrities to help them explore their identity in a safe environment, particularly where real-life sources of support and information are lacking; they were a safe haven.

Sacred

But what if we carry our parasocial relationships with us into adulthood? Until very recently, those who continued to have 'crushes' during adulthood were seen to be suffering from some form of psychopathology or attachment disorder which meant that they needed to turn to untouchable celebrities to plug the inadequacies in their real-life social network. Or, worse still, their behaviour bordered on an obsession which risked tipping over into stalking. But in 2017 Gayle Stever, writing from the basis of evolutionary theory, gave another perspective. She argued that while a proportion of adult fans could be argued to be suffering from some form of psychopathology, there was a solid 80 per cent who gave no indication of mental ill health and seemed to only benefit from the connections they felt to their celebrities of choice. And this group, contrary to earlier prediction, were mostly not forming attachments to celebrities to fill an inadequacy in their social lives. Rather, their parasocial attachments were spoken of as close friends who complemented their social network rather than compensated for a lack of connections in their everyday lives. And because Stever has followed her cohort for nearly three decades, she was able to show that the vast majority had gone on to have successful careers and families. Rather than anomalies or outliers they were functioning and successful members of society.

Why is this the outcome rather than the dire consequences predicted by those who have studied parasocial attachments in the past? The answer lies in the slow pace of evolution and the relatively recent emergence of media, and celebrities, into our lives. While film was invented in the early years of the twentieth century, it only really became available to the masses on a regular basis with the development of television, ownership of which became widespread in the 1950s. From this point onwards, the methods by which we can see, talk and read about our chosen celebrities have only multiplied to the extent that you could now spend twenty-four hours a day seven days a week tuned into their life. But in evolutionary time this is a mere blip on the timeline of human evolution; *Homo sapiens* evolved around a quarter of a million years ago. As a consequence of this, our brains have not evolved to perceive any difference between the human we see standing in front of us and the human on the screen. So we assess and interact with them in the same way, including falling in love. And when it comes to having the attributes of an attractive individual, celebrities tend to have hit the jackpot. They are generally attractive, with abundant resources and, because more so today than ever celebrities remain relevant largely because of the engagement of their fans, they work hard to portray themselves as an appealing friend, potential lover or reliable family member. Lady Gaga, for example, is known as Mother Monster and she refers to her fans as Little Monsters, and has a tattoo to prove it. As Stever states, 'Rather than seeing this tendency to feel a connection to familiar media personae as dysfunctional or potentially pathological, forming such connections could be a natural outgrowth of the evolutionary mandate to become connected to the familiar faces, voices, and personalities of other humans with whom there has been repeated and consistent exposure. Such behavior in human beings is adaptive and insures safety and procreation.' In these connections we find safety, reassurance, support and even validation, enabling us to deal with the real world more robustly. And we select our celebrities in the same way we select our friends or lovers. Bradley Bond's paper exploring parasocial relationships among LGB and heterosexual teens made it clear that the strongest parasocial attachments existed where there was a high degree of interaction, attraction and similarity between adolescent and celebrity – the pre-requisites of a successful relationship in real life.

When I heard the news about Jesy leaving the band it was heart breaking because it felt like I was losing someone dear to me. It was really, really painful. I felt as I did when I had my first break-up, it was such a recognisable feeling. I had a pain in my heart and it happened so quickly. First she was there then the next minute she was gone. I worry we will lose her completely, she will stop using social media. The fear is she is completely gone. **Harmony**, Little Mix Stan

We don't yet have sufficient evidence to declare on whether parasocial relationships are grounded in love, but it is clear that for many they are an important element in their social network and can display the attributes of attachment. Celebrities can be havens of safety and security and provide support or advice during difficult times, particularly where this might be lacking in the real world. And with the advent of social media, maintaining proximity is now easier than ever before with the added possibility of actual connection via a 'like' on your comment or even a 'follow' back. And in a way the attachments we can build with our god or a celebrity show the constant desire we have to make human connections even if we cannot touch them or even see them. For me, this ability to love even in these less concrete environments makes human love even more powerful. The possibilities for love in our lives are truly awe-inspiring.

Up until now, this story of love has been, on balance, one which places love in a positive light. Healthy love is overwhelmingly a good thing and we are lucky to experience it in so many different ways and with so many different beings, including those we can't even see. But the physical and psychological addiction to love which I explored in Chapter 2 has, like all addictions, a dark side. Love *is* central to human experience but because ultimately we rely on it for our health, wellbeing and survival, this reliance can leave us open to exploitation, coercion and abuse. What is love? Love is control.

CHAPTER NINE
CONTROL

● ● ● ●

Earlier this year, 10 people located around the United States were arrested and charged in an organised money laundering scheme. What was strange about the scheme is how the money was obtained in the first place. It wasn't through the trading or trafficking of illegal goods or drugs, but rather cash that was sent by unsuspecting women who thought they were building relationships with the scammers.'

A. J. Dellinger, 25 November 2019, Forbes.com

Ultimately, love is about control. Evolution saw fit to furnish us with a set of reward chemicals which are released when we interact with our lover or our child to bribe us to stick around, cooperate and invest in them for the good of the species. This is a benign control – a control of which we are hardly aware and one which brings many positive benefits to our health and general happiness – but is control none the less. But the addictive nature of these chemicals – and the vulnerability of our psychological need for human physical and mental connection – means that love has also been used as a tool of exploitation, manipulation and abuse. Indeed, in part what may separate human love from the love experienced by other animals is that we can – and some-times do – use love to manipulate and to control. The excerpt from A. J. Dellinger's report in *Forbes* which opens this chapter makes it clear that our need for love means that its exploitation stands alongside those more conventional routes to large sums of money, drug trafficking and stolen goods, as a reliable method of extortion. And this is not small fry. The victims of the reported scam handed over $1.1 million in total, with one poor victim showering her fictitious beloved with over half a million dollars' worth of cash and gifts before the game was up. Overall, romance scammers cost US victims $201 million in 2019 alone, an increase of 40 per cent from 2018. Scammers haunt dating sites with fake profiles and photos and slip into victims' DMs to build a fictitious romantic relationship, full of promises of enduring love but reluc-tant to meet in person, before they start to ask for ever increasing amounts of money. One victim, let's call her Anne, thought she was building a relationship with Stan, an 'intelligent, smart and honest' man, who professed that she was his ' fantasy, love, dream' before conning her out of $200,000; her savings, half her pension and two loans she took out on his behalf. For Anne, love could hardly be described as a positive experience.

In this chapter I want to explore how arguably our greatest and most intense life experience can be used against us, and how it can sometimes also lead us to continue with relationships whose negative consequences act in direct opposition to the instinctive drive to survive. We'll consider jealousy and meet the Dark Triad personalities whose mate-retention techniques lean decidedly towards the negative. We'll ask whether love has a role in rela-tionships where intimate partner violence is the norm, and how the most objectively unpleasant leaders can hold on to power for years, buoyed up by their evangelical supporters. And I want to delve into the centuries-long pursuit for an elixir of love – a chance to gain control of the normally unpredictable pursuit that is catching and keeping love. While in the past love drugs were the stuff of romantic fantasy, today we might be closer than ever before to harnessing love. But is this a Rubicon we really want to cross?

The Anatomy of Jealousy

Love turned to hate is a classic of poetry, of plays, of great literature and they are different sides of the same coin. Very powerful emotions can flip both ways. If love is possessive or needy, then yes, it can be very destructive. **Steve**

If you ask someone whether love has any negative sides, the one they are most likely to land upon is jealousy. As with all emotions, jealousy evolved to promote our survival, the intense and at times aggressive feelings it arouses motivating us to guard our mates and ensure that our genes, rather than those of a love rival, pass on to the next generation. In humans, jealousy exists at three levels: the emotional – dominated by anger, fear and sadness; the cognitive – thoughts around blame, comparisons between oneself and one's rival and plans for revenge; and the behavioural – yelling, the making of accusations, surveillance, distancing and confrontation of the rival. We have all experienced jealousy; in the school yard when our best friend plays with the new girl, at the party when our lover seems just a little too animated in their conversation with a fellow partygoer and sometimes as parents when our toddler seems to prefer our partner to ourselves. And how we react to this perceived betrayal – do we descend into jealousy or re-double our efforts to maintain our relationship? – is influenced both by our gender and our attachment style.

Professor David Buss is one of the pioneers of the anthropology of human mating. It is his groundbreaking thirty-seven-culture study that showed that the sex differences in human mating be-haviour which we encountered in Chapter 2 are universal and that love is a constant wherever you were born or live. In his book *The Evolution of Desire*, he dedicates a whole chapter to the mecha-nisms which enable lovers to stay together, including jealousy. He has spent many years exploring one of the recurring outcomes of studies on jealousy; the stark gender difference. While men and women experience jealousy with the same intensity, *what* triggers a jealous response differs. For men, sexual infidelity leads to the most extreme reactions, while for women emotional infidelity is the biggest trigger. David's book is full of the most wonderful, mostly unpublished, studies which explore this phenomenon. In one study he asked 511 male and female college students about their responses to sexual and emotional jealousy. He presented two different scenarios. One where their partner had sexual in-tercourse with someone else and one where he or she had formed a deep emotional attachment to someone else. The results were clear. For women, the emotional infidelity was the most trigger-ing, with fully 83 per cent responding that this scenario made them the most jealous, compared to men where less than half, 40 per cent, responded that it raised their ire. In contrast, 60 per cent of men found sexual infidelity to be the more jealousy-inducing scenario, compared to 17 per cent of women.

As jealousy is a total body experience – jealousy is often ac-companied by nausea, sleeplessness, tremors, headache and flushing (an evolved response to make sure you are motivated to *do something*) – there is also a sex difference in which types of infidelity lead to the greatest physiological response. Buss and his team hooked sixty men and women up to a range of monitors; electrodes on the forehead to measure frowning, monitors on the first and third fingers to measure sweating and on the thumb to measure heart rate. The two different types of scenario were then introduced – sexual and emotional infidelity – and the monitors got to recording data. For men, sexual infidelity was the most physiologically distressing. In response to this scenario, their heart rates increased on average by five beats per minute – that's the equivalent of three cups of coffee in swift succession. Their skin conductance – that's a measure of sweating – increased by 1.5 micro-siemens as compared to no change from baseline when presented with the emotional infidelity scenario. And their frowning increased; 7.75 microvolt units of contraction to sexual infidelity compared to 1.16 to emotional. It would seem that while sexual infidelity got them worked up, emotional infidelity merited hardly a flicker. In contrast, women found emotional infidelity the most physiologically distressing. Their frown lines deepened in correspondence to 8.12 microvolts of contraction when imagining the emotional-infidelity scenario but only 3.03 units in response to sexual infidelity.

Why does this difference exist? Jealousy is an evolved response to a threat to the stability of a breeding relationship and because what can jeopardise the successful transmission of genes down the generations differs between the sexes, because they bring different resources, so what triggers this response differs between men and women. The biggest threat to a man's reproductive success is the possibility that the child he invests in is not his. As a consequence, sexual infidelity is the biggest threat to him and arouses the largest jealous reaction. In contrast, for a woman the biggest threat to her reproductive success is that she will lose the resources that are critical to the survival of her child and as a consequence emotional infidelity, and the risk that she will have to share if not lose completely the provision and protection provided by her romantic partner, produces the most intense response.

Love can create lots of terribly hateful emotions like jealousy and jealousy is an extraordinarily destructive form of love which only tends to start because you love somebody and you want all of them to yourself and you can't have that. And you get to the point where you can't deal with your partner speaking to another person or admiring another person. **Kate**

Where jealousy is concerned, all of these feelings, thoughts and behaviours are intended to cause one of three different outcomes: to cut off the rival, to prevent our partner's defection or to cut one's losses and move on to the next relationship. But which route we ultimately take is in part down to our romantic attachment style. In their 2017 paper entitled 'Gender differences in romantic jealousy and attachment styles', a group of Turkish researchers, led by Oya Güçlü, explored how our gender and attachment style might influence the way we express our jealousy. Their cohort consisted of eighty-six married, heterosexual couples who had signed up for marriage counselling. Before they spoke to their counsellor, they were asked if they would be interested in being part of the study and, if they were willing, they were asked to complete three measures: one relating to sociodemographics, one related to romantic jealousy and one related to adult attachment. The measure for romantic jealousy was The Romantic Jealousy Questionnaire, which presents scenarios relating to both sexual and emotional infidelity, and measures five different subscales of jealousy: jealousy level, response to jealousy (emotional, cognitive and behavioural), coping with jealousy, effects of jealousy and reasons for jealousy.

What Oya and her team found was that while women reported higher emotional and cognitive jealousy than men, the lack of a difference between the genders in behavioural jealousy, of enacting the jealousy, suggested that women were consciously controlling the expression of their jealousy. When it came to mechanisms to cope with jealousy, the team identified four different dimensions: exit – including threats to leave; loyalty – both actively and passively hoping for improvement; neglect – both actively and passively allowing the relationship to die; and voice – actively working to save the relationship by talking. While men and women tended to use these coping mechanisms equally, women did tend to rely on the mechanism of loyalty – sticking around – to a greater extent than men, reflecting the findings of other studies which showed that women tended to use more constructive coping mechanisms than men when confronted with infidelity and jealousy.

I think if people have unhealthy or selfish attachment styles even if they don't mean to, even if it is subconscious, they can do the people they love great unkindness but I am not sure if that . . . I mean, I guess it is love. It's not healthy love . . . **Jo**

Finally, the big question: are some attachment styles more jealous than others? Previous studies had suggested that those with an anxious attachment profile tend to experience more emotional and behavioural jealousy – and display more anger and controlling behaviours – than those who are avoidant, where cognitive jealousy comes to the fore. Remember that anxious attachment style is associated with high levels of anxiety about abandonment. However, in this study the team found that only anxious attachment style seemed to have an influence on the expression of jealousy, with anxious people tending to score higher on all three dimensions of jealousy, indicating that these individuals may be prone to obsessional thinking as well as anger and controlling behaviour. In contrast, secure individuals were the only group who experienced any positive effects of jealousy as they used their emotional and cognitive responses to formulate behaviours aimed at resolving the issue and maintaining the relationship in the long term. We tend to view jealousy as overwhelmingly negative, but this finding shows us that it can be a positive and effective maintenance tool, as evolution intended, as long as we keep its extreme outcomes in check. But the question arises: what happens when jealousy gets out of hand?

The Dark Triad

Love can be negative if you love someone and they don't feel the same way. In my experience, you can go too far in trying to win someone's love for you, and sacrifice yourself – your physical and mental health – to do that. Or you accept compromised behaviour on their part that damages you because of their lack of ability to show you the love you want or need. **Stella**

I want to introduce you to the Dark Triad. This is a set of personality traits which seem to significantly influence the types of mate-retention behaviours, of which jealousy is one, someone will use. The Dark Triad consists of Machiavellianism, psychopathy and narcissism, and where someone is in possession of at least one of these three they are, it is suggested, much more likely to use mate-retention tactics which inflict a cost on their partner – violence, manipulation, aggression – than ones which confer a benefit. While they are distinct psychological concepts, all three tend to produce individuals who are callous, manipulative and exploitative in nature, and who accrue benefit at a cost to others. Machiavellianism, named after Machiavelli, the author of *The Prince*, describes individuals with a lack of empathy but an ability to utilise advanced mentalising skills to emotionally and behaviourally manipulate others to achieve their own ends. Psychopathy is likewise a lack of empathy, a tendency towards anti-social behaviour and manipulation and a liking for sporting with the feelings of others for pleasure. Narcissists exhibit extreme self-love, tend to be entirely self-centred and will devalue others to improve their own standing. Self-glorification is the goal. However, while these individuals sound like the partner from hell, their persistence within our species suggests that, while being costly to others, their ability to exhibit these personality traits confers a survival advantage on those who possess them. They are incredibly good at looking after number one.

In 2018 psychologists Razieh Chegeni, Roshanak Khodabakhsh Pirkalani and Gholamreza Dehshiri set out to explore how the Dark Triad influenced the use of beneficial and costly mate-retention behaviours in an Iranian cohort. Was it more likely that those in possession of the Dark Triad were more prone to coerce and abuse? Beneficial mate-retention behaviours are those which tend to increase relationship satisfaction, for example, gift giving or helping out at a cost to yourself, while cost-inflicting mate-retention behaviours confer psychological, emotional and physiological cost upon the recipient. So this could be actual physical injury or damage to mental health, for example gaslighting. Razieh and her colleagues recruited 205 participants, 54 per cent female, with a mean age of thirty-two and an average marriage length of six and a half years. They asked them to complete two self-report measures. The first was the Mate Retention Inventory (Short-Form) devised by David Buss. This assesses nineteen different mate-retention techniques using thirty-eight questions. It asks participants to record the frequency with which they have used a particular behaviour over the last year. Example questions include 'snooped through my partner's personal belongings' (vigilance), 'talked to another man/woman at a party to make my partner jealous' (jealousy induction), 'bought my partner an expensive gift' (resource display) and 'slapped a man who made a pass at my partner' (violence against rivals). Alongside this, participants were asked to answer the fabulously named 'Dirty Dozen'. This measure uses twelve statements to assess the extent to which a person possesses the Dark Triad; Machiavellianism, psychopathology and narcissism. In this case, example statements included 'I tend to want others to admire me' (narcissism), 'I tend to be unconcerned with the morality of my actions' (psychopathy) and 'I tend to exploit others to my own end' (Machiavellianism), with participants asked to indicate the extent to which these statements applied to them. Chegeni and her team then set out to explore whether having a Dark Triad personality altered mate-retention behaviours.

Results showed that, in the first instance, men used more mate-retention behaviours than women. However, possession of a Dark Triad personality correlated with both cost-inflicting *and* benefit-inducing mate-retention behaviours although the relationship with cost-inflicting behaviours was both stronger and of more significance. This suggests that while Dark Triad personalities tend to rely to a greater extent on cost-inflicting mate-retention behaviours – aggression, manipulation and exploitation – benefit-inducing behaviours are also employed at times to ensure that their partner does not stray. This oscillation in behaviours is something that is often seen in abusive relationships. It keeps the victim on their toes and introduces doubt into their mind – this is gaslighting – regarding whether or not their partner really is emotionally, psychologically or sexually abusive, making them more likely to stay under their control.

The Darker Side of Emotional Intelligence

Up until this point, this book has always presented emotional and social intelligence as good traits to possess. Previous chapters have explained how giving your child a stable and secure upbringing leads to the development of empathy, and a set of prosocial skills which will enable them to have healthy and happy lives going forward, full of functioning and beneficial relationships. They will exhibit high levels of social and emotional intelligence. But the fact that the Dark Triad traits, in particular Machiavellianism and psychopathy, are based on the ability to manipulate others suggests that strong mentalising skills – the basis of emotional intelligence – are not always used in a positive way. In their paper 'Is there a dark intelligence? Emotional intelligence is used by dark personalities to emotionally manipulate others', a team of psychologists, led by Ursa Nagler from Leopold-Franzens University in Austria, questioned whether the combination of high emotional intelligence and a Dark Triad personality led to an increased risk of the emotional manipulation of others. They asked whether socio-emotional intelligence (SEI) inherently has a dark side which we have been slow to recognise, or whether this dark side rears its ugly head only when used by a Dark Triad personality.

Ursa and her team recruited 438 women and 138 men to take part in their study. They administered a set of questions which measured socio-emotional intelligence, narcissism, Machiavellianism, psychopathy and emotional manipulation. Their results showed that in the first instance, aspects of SEI were related to elements of psychopathy, Machiavellianism and narcissism. Narcissism and psychopathy showed a positive relationship with socio-emotional expressivity and control and emotional manipulation, but a negative relationship to socio-emotional sensitivity. So this group use advanced SEI to have a deep understanding of their own experience, and use this to manipulate others to gain what they want, but have little interest in the experience of the other. For Machiavellianism, the relationships were largely negative, with the exception of emotional control and manipulation. These findings suggest that psychopathy and narcissism, in particular, are linked to high socio-emotional intelligence including control of one's own and other's emotions, a key aspect of emotional manipulation. Secondly, the link between socio-emotional control and emotional manipulation is mediated by the extent to which a person shows aspects of narcissism and psychopathy in their character. The greater the contribution these personality traits make to your character, the more likely you are to use your high socio-emotional intelligence for darker means.

Abusive Love

But it's blind love. It's difficult to understand that a person may hit you numerous times but you still go back to them. From the outside you may conclude by thinking that maybe he really does love her ... Anonymous victim of domestic violence (Singh et al., 2017:22)

A recent World Health Organisation report covering the experiences of 24,000 women from ten countries from Bangladesh to Ethiopia, Serbia to Japan, reported that 61 per cent of women had been physically abused, while an enormous 75 per cent had been emotionally abused, by a romantic partner. In the US the Center for Disease Control and Prevention (CDC) questioned both men and women about the incidences of domestic violence they had experienced in their lifetime. Looking at severe physical abuse alone – that means being punched, slammed, kicked, burned, choked, beaten or being attacked with a weapon – one in five women and one in seven men reported at least one incidence in their lifetime. If we consider emotional abuse, then the statistics for men and women are identical – 48 per cent of both sexes have experienced psychological aggression. And the real extent of such abuse is likely to be far greater, as fear and cultural factors prevent the reporting of some incidences.

What you allow to happen because you love this person and you step back and with the value of hindsight. I mean my ex-wife pouring hot coffee over her dad's head, not really an acceptable thing to do. My ex-wife dragging her own mother by her hair out of the house, not really acceptable. But I did it . . . **Colin**, survivor of domestic abuse

Those who have endured violence or abuse often speak of the power of enduring love, its ability to overcome all despite strong evidence to the contrary. However, even when the victims acknowledge that their love is a long way from the 'norm', how can we be sure that what they feel is love? Certainly, in carrying out interviews for this book I have asked many interviewees the question: 'Can love ever be negative?' Some respond in the affirmative

– love can lead us to be abused and manipulated, can make us subordinate our fundamental needs to another, or, where parents and children are concerned, can smother a child's practical and emotional development and prevent their independence, such is our need to keep them close. But an equally powerful group have answered 'no', because if love has negative consequences it isn't love, the term is merely being used as cover for damaging behaviours. So is there a place for love in these relationships?

She was keen not to let me out of the house in the end. She asserted she feared for my state of mind. It got to a point she was trying to stop me going out and I just made a belt for the door into the car and while I started driving away she opens the passenger door and jumps in and is reaching across and grabbing the steering wheel, probably trying to grab the keys. **Justin**, survivor of domestic abuse

In their 2013 study, three mental-health nurses from West Virginia University in the US, led by Marilyn Smith, explored what love meant to nineteen women who were experiencing, or had experienced, domestic violence. In the first instance they defined for their readers what IPV – intimate partner violence – can constitute. It is harrowing reading and includes, but is not limited to, 'slapping, intimidation, shaming, forced intercourse, isolation, monitoring behaviors, restricting access to healthcare, opposing or interfering with school or employment and making decisions concerning contraception, pregnancy and elective abortion'. They held focus groups and one-to-one interviews to explore the answers to a set of questions which asked the women to reflect upon what they believed love between a man and women was and what it involved, how they came to learn about this model of love, how their relationship compared to this model, how important it was to remain in their current relationship and why it was important to be in a loving relationship. At the end of the session, women were asked if there was anything else pertaining to their experience which they wanted to talk about. Following the sessions, all the women were offered further support.

Marilyn and her team analysed the transcripts of the focus groups and interviews to identify any common themes. In the first instance, the women spoke about what love was not: being hurt and fearful, being controlled and having a lack of trust and a lack of support or concern for their welfare. Here are a few quotes to illustrate these findings:

When he was mean, he would whip me and the kids . . . He would come in and tear the furniture up or turn the refrigerator and table upside down. He would lie in bed with the four boys and me and he would shoot a gun between us. He would go, 'One, two, three, four,' and he would shoot between us. He said he wasn't trying to hit us, he was just practising. To live in fear is not good.

He wouldn't let me dye my hair. He never wanted me going anywhere. He would always say, 'Don't tell me the N word.' And that was 'NO'.

Despite acknowledging that these behaviours were not love, and understanding what a loving relationship should look like – respect and understanding, communication, support and encouragement and commitment, loyalty and trust – the women expressed that they felt an attachment to, and loved, their partner and it was for this reason overall that they stayed, sometimes for years. Some of the women felt a need to nurture and protect their partner or be nurtured and protected by them. For others, they were hopeful that their partner would change and they were ready to forgive the man they loved. A fear of being alone or of beginning a new relationship which could be even worse were among the reasons to stay. And strong ideas of what a family should be – nuclear rather than single parent – made some women reluctant to leave.

I thought, Well, there is some good there somewhere. Maybe, I am going to bring it out. The more that I would think like that, it seemed like to me, the deeper I felt like I loved him.

I would think about hoping for a change because they said that if you love somebody deep enough, they will begin to change.

And my own interviews with male survivors of domestic abuse make it clear that as with the female survivors in Marilyn's study 'white-knight syndrome' can cause victims to remain in abusive relationships much longer than is healthy for them:

I thought I could save her from her childhood trauma. If I stuck by her long enough despite everything she was throwing at me, one day she would realise I wasn't like the guys who had abused her and she would realise I was a good guy and I did love her and wanted to be with her. But the more I gave emotionally, physically the more she took and the worse the abuse got. **Mark**, domestic-abuse survivor

It is not only love for the partner which makes a victim stay with the perpetrator but their love for their children and the fear that, if they should leave, their child will be left to face the situation alone. During the writing of this book I have spoken to male survivors of domestic abuse, a perspective we rarely hear, and for those with children the prevailing belief both culturally and legally within the UK that a child is better off with their mother means that losing contact with their child if they leave, and as a consequence being unable to protect their child, is a very real prospect.

. . . the real strength of love that I had wasn't for my ex-wife, it was for my daughter who was a baby. That was what kept me there . . . she'd use the baby as emotional blackmail to either get me out of the house or get me out of the car and leave me stranded seventy miles in the middle of nowhere from home . . . and so . . . but that love for my daughter was being exploited by her as a means for controlling me. That went on for the best part of ten years. **Paul**, domestic-abuse survivor

The Darker Side of the Fairytale

Eventually it reached the point where even what I was eating was controlled. I had to hide food in the boot of my car because I worked from home. When she eventually went to work I could eat something I wanted to eat. Every household decision was hers and one of the defining moments was when she turned around to me one day and said why are you giving me your opinion? You know it counts for nothing. **Mark**, domestic-abuse survivor

For many people who are abused, the exclusivity that their partner insists upon, and the jealousy he or she exhibits, can be perceived as love in the initial stages of the relationship. By the time it becomes apparent that this exclusivity is underpinned by an imbalance of power and a desire to control every aspect of their life, the victim is often dehumanised and isolated to such an extent that it is difficult to find the confidence and support to leave. Some victims are filled with guilt as they contemplate breaking the exclusive bond with their partner or report needing such a close bond to make them feel complete. But is it also a possibility that our culturally influenced ideas about romantic love itself are to blame when women and men stay and endure intimate partner violence (IPV)?

I think it's all due to love. No matter what a person can do to you, if you love him you just love him. You can be angry for a while but it will pass and you'll forgive each other and move on. Girls believe in changing people, so they stay with that hope but they don't change. **Anonymous** victim of domestic violence (Singh et al., 2017:28).

South Africa has one of the highest rates of IPV against women in the world and rates as one of the leading countries when it comes to death at the hands of a partner. In their paper, published in 2017 in *Agenda: Empowering women for gender equity*, Professor Shakila Singh and secondary-school teacher Thembeka Myende explored the role of resilience in female students at risk of IPV, which is prevalent at a high rate on South African university campuses. The paper ranges widely over the role for resilience in resisting and surviving IPV but what is of interest to me is the fifteen women's ideas about how our cultural ideas about romantic love have a role to play in trapping women (and in all likelihood men) in relationships with IPV. They point to the romantic idea that love overcomes all obstacles – family objections, geographical difference, illness and poverty – and must be maintained at all costs even when abuse makes these costs life-threateningly high. Or the idea that love is about losing control, being swept off your feet – the dashing prince rescuing you and carrying you away to his castle. Having no say in who you fall for even if they turn out to be an abuser. Or that lovers protect each other, fight for each other to the end even if the person who is being protected, usually from the authorities, is an abuser. Or the belief, underpinned to a certain extent by science, that love is blind and we are incapable of seeing our partner's faults despite them often being glaringly obvious to anyone outside the relationship. It is these cultural ideas about romantic love, the women argue, that lead to the erosion of a woman's power to avoid entirely or to leave an abusive partner.

To begin with, you were under the impression you took this step to actually get married to somebody, that it was what you thought it would be but it is very easy to pull the wool over people's eyes sometimes or also sometimes you get those rose-tinted spectacles, don't you, and you want it to be better than it actually is even if somebody is actually saying to you, 'Hang on, something isn't quite right.' You want them to be wrong. **Colin**, survivor of domestic abuse

It is difficult to speak to someone else's experience and I hope it is clear by now that a significant part of love is subjective, unique to the individual. For those who sit on the outside of these relationships it can be difficult to understand how this could possibly be love, so far is it from the stuff of romantic movies and fairytales. There is little to no science about love within abusive relationships, it is difficult ethically to place such vulnerable people in a scanner or subject them to intrusive genetic and neurochemical scrutiny. We are lucky to have the voices of those brave few, both men and women, who have been prepared to speak to us about their relationships but who ultimately, as we all do, have the right to define their love as they want, however difficult the rest of us might find it to understand and whatever the scanner screen may say.

'Yes, We Can!'

While in the home the abusive partner or parent controls the spouse or child, in a broader societal context leaders have the power to lead, or control, whole nations or religions. The most successful and charismatic leaders employ their skills of eloquent communication and persuasion as well as sheer force of personality to lead their followers. Leaders can be forces for huge and positive change, enabling individuals to achieve the unimaginable and nations to triumph over adversity – think of Gandhi, Nelson Mandela or Barack Obama. Historically, charisma was the stuff of royalty or religious leaders – those who had a direct connection to the divine – and those who followed their lead were promised great riches and even to share in some of the supernatural powers of their idol. Even today, our susceptibility to the draw of charismatic leaders shows no sign of abating. The fact that their powers have endured points to the evolutionary advantage that these characters confer upon the groups they lead. Evolutionary psychologist Mark van Vugt has spent a career observing the behaviour of charismatic leaders both in politics and business and he argues that we need charismatic leaders because they are uniquely capable of organising large-scale coordination of people when society requires this. As a consequence, such leaders often rise to power when the people demand change. Further, they have attributes, indicators, that clearly advertise to potential followers that they are the person for the job. Indeed, Mark argues that there is a 'leadership index' which we have evolved to rapidly calculate, a bit like the algorithm for selecting a suitable date we encountered in Chapter 2, to identify who is an attractive leadership candidate. So there are key physical cues associated with charisma: height and strength, indicating formidability; facial attractiveness (the more likely to attract attention); and fluid speech and movement, indicating health and high energy. They must also bring a substantial social network to bear, indicating that they have the ability to form functioning coalitions and implying the welcoming arms of an already established 'family'. But the context of the leader's rise to power is also important; some cues which are relevant in conflict are not relevant or helpful when change is required during peacetime. It takes a different sort of personality to engender a group's sense of unique identity against a foe from one who needs to seek conciliation and cooperation. Churchill's considerable bulk and role as an experienced politician was valuable in depicting him as a formidable foe during World War II, whereas Barack Obama's call for collective action during his presidential campaign, 'Yes we can!', or JFK's words spoken during his inauguration speech in 1961 – 'Ask not what your country can do for you but what you can do for your country' – ask for cooperation and self-sacrifice as the drivers for change in a peacetime society. Here, both Obama's and JFK's youth and vitality were critical in representing a new way of living. However, regardless of the context, charismatic leaders rely on a set of behaviours, both their own and those of their followers, which cement the bond between leader and follower. They must be capable of attracting attention, of having minute control over their expressions, body language and voice pitch and pace, of arousing powerful emotions, of giving virtuoso rhetorical performances, of producing performances which are designed to engage all the senses and to exhibit the ability to instil a sense of urgency and a need for collective change.

Most importantly, they must appeal to their prospective followers *in public*. The grandstanding public speech is key because it is
the presence of others which makes a leader's message so attractive and powerful. This is in part because when appealing to your
shared values they rely on the presence of others to reassure you
that they share these values and you are a member of a group; a
shortcut to trust. But *what* the crowd are compelled to do is also
critical, and here we find the closest parallel with human dyadic
love. Because one marker that political, military and religious ral-
lies share is the power of synchrony and the use of synchronous,
endorphin-inducing behaviours by charismatic leaders to bond
their followers and increase the love they receive from them. For
Barack Obama, it was the repeated, synchronous chanting of 'Yes,
we can!'; for armies, it is marching in step, for religious leaders it
is chanting, singing and dancing. As we know from Chapter 2, all
these behaviours when carried out alone will produce hits of beta-
endorphin, the neurochemical of long-term love, but when we do
them *together and in synchrony* then the quantities of endorphin
released increase exponentially, the feeling of euphoria is ramped
up and our addiction is swiftly established. By encouraging their
supporters to carry out these endorphin-rich behaviours, charismatic leaders are ensuring that they become addicted not only
to each other, but to the very leader who has brought them all
together in the first place.

You may ask why charismatic leaders have a place in a chapter focusing on the darker sides of love? But just as a new lover
is blind to the faults of their new partner, so recent research by
Danish psychologist Uffe Schjoedt has shown that just by believing that we are interacting with a charismatic individual is enough
to lead to downregulation of the part of the brain responsible for
spotting errors or incongruities in what we hear or see. Indeed,

we may be so blinded by their presence that we allow them to control the narrative of our meeting, even if their version flies in the face of the facts (that's former President Trump's 'alternative facts'). When it comes to leaders of positive change this may be of minor concern, but there are numerous examples of our blindness to a charismatic leader's true intentions leading us, our country and even our world into a very dark place.

The Making of a Despot

I imagine there are few in the world who have not seen the film of the 1935 Nuremberg rally at which Adolf Hitler used all of his rhetorical might to preach a message of Aryan supremacy and deep-seated hatred for those he conceived to be 'lesser races' to the massed ranks of his chanting, Sieg-Heiling and marching followers. Even though it is in black and white, it is an overwhelming vision of noise, movement, strong symbols of membership and excess of monumental buildings and people; an assault upon the senses which is hard to forget. Bring it right up to date and focus upon the nationalist leaders of today – Trump, Bolsonaro or Erdoğan – and their use of the mass crowd to spread their message alongside synchronous chanting, fist pumping and the clear identification of a common enemy makes for a chillingly familiar spectacle. For as with dyadic love, although charismatic leadership has evolved and endured because on balance it is beneficial, there is always the risk that it will be exploited for the benefit of the individual leader at a cost to their followers.

In the same way that our need for individual love can be exploited to abuse, so our desire to belong and be led can be exploited by hugely attractive individuals for their own end. The people of North Korea profess their love – seemingly of their own volition – for the Supreme Leader, sporting state-approved hairstyles which reflect those of Kim Jong-Un and his wife and placing a cross next to his name, the only one on the ballot paper, in the five-yearly election. For Donald Trump, his use of campaign-rally-style meetings throughout his presidency, rather than just in election year, showed his understanding of the power of the supportive, baying crowd. His chosen enemy, the media, were loudly and synchronously decried, 'CNN sucks' being a popular chant, and his bright red 'Make America Great Again' caps a distinctive symbol around which to congregate. Turn to religion and the followers of the leader of the LA-based Hill Song Church, Carl Lentz,[1] willingly donate money to his church on the promise that they are pleasing God. With 900,000 followers on Twitter and 750,000 devotees of his family's personal Instagram account, a plethora of celebrity followers including Justin Bieber and Nick Jonas whose presence make this family a highly attractive one to join, and a turnover from donations of $100 million in one year, Lentz's leadership is certainly working. It is no coincidence that, in order to win people over, Jong-Un, Trump and Lentz all employ techniques such as marching, singing, exercising and touching – the natural generators of the neurochemicals which will secure and reward the bond developing between themselves and their disciples. Remember that synchronous group actions such as these lead to an exponential ramping-up of beta-endorphin, leading to intense and addictive bonds. Indeed, view a YouTube video of the Sunday service at Hill Song Church and you will be assailed by a

---

1 In 2020 Lentz and his wife were fired from Hill Song Church for moral failures; he admitted to having an affair and there are allegations of sexual abuse and inappropriate behaviour.

vision of thumping music, flashing lights and cheering, dancing and singing people in their hundreds. It is a frenzy of worship and the perfect environment for a massive endorphin release leading to feelings of euphoria, love and, most critically, unquestioning loyalty to the beloved leader.

Ultimately despots desire control and are past masters at asserting that control by the behaviours and beliefs they induce in their followers. While they take their quest for control and dominance to the extreme, as the plethora of self-help guides, advice columns and therapeutic services surrounding romantic love attest, we all have a tendency to want to quell the unpredictability of love to some extent. Indeed, many of the people who attend my talks on the science of love do so with the hope that I might give them some guidance or even reveal to them the formula for love. But here's something for you to consider. If there was a drug that claimed to be able to control love, would you take it?

The Elusive Elixir of Love

Well, for me I guess [taking ecstasy] is like the ability to let the guard down, to take away those blocks and the walls and just having the ability to be completely open to anything and anyone. **MoJ** (Leneghan, 2013:352)

For at least 2000 years, humans have sought to control love – induce it or wipe away its pain – through potions and elixirs. The Persian polymath Avicenna who lived at the end of the tenth century provided a treatment for lovesickness in his writings on pathology under the title 'On Obsession: A Melancholic Disease'. In Shakespeare's *A Midsummer Night's Dream*, Oberon, king of the fairies, administers a love potion to his sleeping wife Queen Titania which has the effect of making her fall in love with the first being she sees on waking; in this case, an ass headed man called Bottom. And Donizetti's opera *L'Elisir d'Amore* draws inspiration from the legend of Tristan and Isolde where Tristan relies upon a love potion to capture Isolde's heart. Even today, in a time of apparent scientific objectivity, type in 'love potions' to Google and the top three questions are 'How do you make a love potion?', 'Do love potions actually work?' And 'Are love potions dangerous?'.

Evidently we yearn for some control over the apparently uncontrollable phenomenon that is love, and there are certainly things I could tell you to do which might increase your chances of catching love or coping with heartbreak. The knowledge that we now have about the behaviours which cause the release of bonding hormones means that if you arrange a date full of endorphin-inducing activity – ballroom dancing or a comedy club are good bets – then you increase the chances that the object of your desire will fall for you. And as breaking up leads to a severe case of cold turkey, due to the rapid descent of beta-endorphin, oxytocin and dopamine levels, having a massage, singing along to the car radio at the top of your voice or going for a run will top up your supplies of these key bonding chemicals, while a good chocolate-eating session will restore your dopamine levels. But for the first time, as our knowledge of the neuroscience and physiology of love continues to grow, the possibility of an elixir that actually works is tantalisingly close. This presents an opportunity to those who wish to make money from the basic human need to love and be loved. Whoever harnesses love, whoever can control it, is set to reap vast monetary rewards, so desperate are we to ensure we have love in our lives.

Time to Spritz and Go

Recruiting subjects for a scientific study is always a nail-biting affair. Will we get sufficient numbers with the right attributes and will they actually turn up? Will they be willing to be poked, prodded, injected and scanned in return for the paltry financial sums, or much-vaunted Amazon voucher, which represents the maximum reward we are allowed to offer by the powers that be in ethics? Will they stick the course or will they quit after experiment two, rendering all their data unusable and condemning the researcher to many sleepless nights? When it comes to my personal experimental history, I have asked people to offer up their blood and their spit, to sit in front of a computer screen while endless questions and scenarios are presented to them, to share their deepest thoughts and experiences with me and my voice recorder and to willingly endure the noise and claustrophobia of the fMRI scanner or, arguably worse, the intravenous radiation of the PET scan, all in the name of furthering scientific knowledge.

In comparison, the request that they be squirted up the nose with oxytocin sounds quite mild. Studies that have employed this technique have allowed us to understand the power that oxytocin has to mould people's emotions and prosocial behaviours. Numerous studies have shown that receiving oxytocin increases empathy and trust and motivates people to cooperate, even when the guy on the other side is a stranger. In parents, a squirt of oxytocin increases sex-specific parenting behaviours – play in fathers and nurturing in mothers – allowing us to come to the sometimes controversial conclusion that dads and mums have evolved to be different. Research utilising synthetic oxytocin is ongoing, as we work to understand how the prosocial aspects of oxytocin treatment can be used to benefit those who struggle with social interaction and cognition, such as those with social anxiety, autism and borderline personality disorder. But did you know the possibility now exists that all of this research can be put to good use to help you up your game in the dating market? You could add a squirt of oxytocin to your pre-going-out ritual of a couple of glasses of prosecco and a dance around your bedroom, before heading into the dating bearpit that is a local club on a Saturday night with your head held high?

Let me introduce you to OxyLuv. A product, available for next-day delivery from Amazon and eBay, which claims to reduce social anxiety, make orgasms more powerful and potentially be a helpful aid to increase your confidence during the search for a mate (as a side note here, it was initially marketed as a pheromone which made you irresistible to other people. One sniff and they fell at your feet!) And as we know from Chapter 2, this is certainly the impact that endogenous oxytocin has upon us in the first few moments of attraction. Remember its key role is in quietening the fear centre of our brain, the amygdala. Should you wish to encourage your amygdala to pipe down, and are willing to part with your pounds, euros or dollars, you will receive a small blue bottle and the recommendation to squirt twice up the nose for a noticeable effect. And as the 38 per cent of consumers who gave it a five-star rating attest, it appears to work for some people.

So is this the elixir of love that we have been searching for for centuries? Well, no. It is certainly the case that a squirt up the nose works well for some, although most of the reviews on Amazon attest to its power to encourage milk let-down in breastfeeding mothers or be an effective (although I must stress untested and unapproved) treatment for autism, rather than its suggested power to ease social anxiety. Indeed, the 25 per cent who gave it a one-star review attest to the fact that it doesn't work for everyone. As one reviewer puts it, '[OxyLuv] just made me tired and emotionally crotchety like an oestrogen dominant bag of self-pity'. Oh dear.

The issue is that as oxytocin research moves beyond the initial exploratory studies and our questions become more nuanced, it is becoming clear that the impact that exogenous oxytocin has on you is highly context-dependent and individual. So while for some it is the elixir of love – increasing empathy and trust and motivating cooperation – for others, it can have distinctly negative effects. For those of us who study the neuroscience of love, this is unsurprising. The neurochemistry that prompts us to form and maintain our relationships is incredibly finely balanced, we still do not have a clear picture of what interacts with what, which chemicals complement each other and which are antagonistic. Add an external form of one of these chemicals to the mix and everything is pushed off-kilter and the outcome can be unpredictable. In their 2011 review of the outcome of experimental use of oxytocin in prosocial tasks, psychiatrist Jennifer Bartz and her team concluded that of the over thirty studies they considered, 43 per cent indicated no significant impact of oxytocin on prosocial behaviour, 63 per cent reported moderating effects of individual traits or the context of administration – so who received the drug and in what circumstances had an impact on its effectiveness – and a significant minority reported *anti-social* effects. Of these anti-social effects, the most striking was that while exogenous oxytocin can increase your prosociality to people who you conceive of as being part of your in-group, it can have the exact opposite impact if you believe the target of your attention is a member of an out-group. Not a good result if catching a date is your goal.

In 2010 Dutch psychologists Carsten De Dreu, Lindred Greer, Gerben Van Kleef, Shaul Shalvi and Michel Handgraaf, bucking the trend in oxytocin research, reported on the outcome of their experiments exploring the darker side of oxytocin and, in particular, its role in promoting human ethnocentrism. Ethnocentrism underpins in-group cooperation by encouraging members to view their group as superior to other groups, thus cementing their loyalty to the group and their reliability as a cooperative partner. In some cases, as well as a positive view of one's own group, ethnocentrism can lead to derogation of other groups. Carsten and his team employed a set of five experiments to robustly explore how oxytocin might encourage ethnocentrism and by association negative thoughts and behaviours about out-groups. In the first three they employed word-association paradigms to investigate how people thought about members of their own group and members of other groups. For example, in experiment three they utilised a task based upon the idea of 'infrahumanisation', which basically means we ascribe more uniquely human behaviours or emotions to in-group members than out-group members. Here, it was the possession of secondary emotions, such as delight and embarrassment, which are thought to require higher cognitive ability and be uniquely human (whether they are or not is a debate for another book).

Sixty-six Dutch males were given either oxytocin or a placebo and asked to decide whether a 'typical' Dutch person (designated the 'in-group') or a 'typical' Muslim person (designated the 'out-group')[2] could possess a particular emotion. In all, three positive and three negative primary emotions were presented and three

2   I'm aware these designations are problematic. After all, there are Dutch Muslims and where do they sit?

positive and three negative secondary emotions. Results showed that where the participants were given oxytocin, but not the placebo, they ascribed secondary emotions, negative and positive, preferentially to the Dutch individual as compared to the Muslim, showing in-group favouritism.

Their last two experiments focused upon classic moral-dilemma scenarios. These are regularly used by psychologists to assess the relative value of different people to the participant. One of the moral dilemmas focuses on a runaway trolley bus which is at risk of killing an individual who is stuck on its tracks. The participant can choose to save the target individual by switching the trolley onto another track but the consequence of this would be the deaths of the ten unnamed people on the trolley bus. Or they can choose not to switch the tracks, killing the target individual but saving the ten unnamed occupants. In this set of studies the target individual either had a Dutch name, an Arab name or a German name. Results showed that participants who had taken oxytocin preferentially chose not to sacrifice the target individual with the Dutch name as compared to the target with the Arab or German name.

What do all these results mean? That oxytocin does appear to cause intergroup bias either as a result of in-group favouritism and/or out-group derogation. So while, if you are focusing on members of your group (and this can be a flexible concept depending upon the identification of that group; family, fans of a football team, a whole country), oxytocin will generally lead to the display of increased trust, empathy and cooperation which we expect, where the focus is an out-group member the exact opposite can occur: the promotion of racism, bigotry and aggression. Indeed, a recently published study by Hejing Zhang utilising a simulated intergroup conflict found that those who had received oxytocin did work more cooperatively with their in-group members during the attack, but doubled down on their aggressive tactics towards the out-group which they sought to defeat. And if we unconsciously view our potential dates in the nightclub as a member of an out-group – as strangers quite possibly are – then we can argue that oxytocin might not be the reliable elixir of love that it is promoted to be.

And It's Back to the Nineties

You can have the same conversation straight, and people don't connect. When they have the same conversation fucked up, and suddenly ... you are best friends for life ... MoJ (Leneghan, 2013: 351).

Whereas oxytocin is clearly associated with the development and maintenance of close human bonds, MDMA, or ecstasy, is best known as a drug of the rave scene. As a consequence, it has tended to garner more negative headlines than positive over the years, having been implicated in the deaths, often at a young age, of recreational users. However, talk to MDMA users about their experiences or read accounts of the rave scene and one statement is repeated again and again. As well as inducing feelings of immense energy and euphoria, users of MDMA report intense feelings of connectedness and love for their fellow dancers. As a consequence, researchers and drugs companies have begun to explore whether there is a role for MDMA in the treatment of a range of psychological disorders, including PTSD, and in conditions which express themselves in the social domain such as autism. And arguments for its use as a therapy during couples' counselling suggest the potential also exists that this is the long-hoped-for elixir of love.

However, as with oxytocin, it is becoming clear that with MDMA context is everything. While it can bestow heightened social abilities and sensations of deep love with one hand, it can as quickly take them away with the other. In their 2010 article entitled 'Is ecstasy an "empathogen"?' pharmacologists Gillinder Bedi, David Hyman and Harriet de Wit explored the impact of ecstasy use on empathy in a group of twenty-one users. Over four sessions, participants were administered with either 0.7 mg/kg or 1.5 mg/kg of MDMA, methamphetamine (to check whether any affects were merely related to the ingestion of an amphetamine) or placebo. They were asked to complete two self-report measures of subjective experience, the Visual Analogue Scale (VAS) and the Profile of Mood States (POMS), to assess changes in sociability, loving, loneliness and friendliness both before and after administration. In addition, to objectively assess changes in empathy, they were asked to identify emotions from two facial-identification tasks, one which presented full faces which differed in emotional expressions and one which only presented the eye area (the famous Reading the Mind in the Eyes test), and one vocal task where actors were asked to read the same phrase, 'I'll be back later', in an angry, sad, happy or fearful tone.

Results showed that the 1.5 mg/kg dose of MDMA in particular had significant impacts on subjective feelings, increasing the sense of loving feelings and friendliness. However, MDMA did not objectively increase the ability to correctly assess emotions but rather decreased the ability for participants to identify negative emotions, particularly fear. As a consequence, the researchers concluded that MDMA does not increase empathy but does increase the motivation to be sociable.

So is MDMA really the wonder drug that it has been proposed to be? A recent replication of Bedi's study, a product of the London-based team led by Professor David Nutt (the UK government's ex-Chief Drug Adviser) which explored cooperation and trust as well as empathy, again concluded that MDMA did not significantly improve these objectively measured aspects of pro-sociality as compared to placebo, although subjective measures of closeness, empathy and compassion again improved. However, as with oxytocin, context appears to be a factor in how effective MDMA is as a prosocial drug. Nutt's team pointed out that the lab setting itself may be a factor in dampening prosocial feelings and behaviours; the average university science lab is not a particularly warm or welcoming place. It may be that, should it be possible to circumvent the health and safety laws, carrying out these experiments in the home or even the club would yield significantly different outcomes. In addition, all of the objective tests presented to the participants in both studies were displayed on a computer screen. Maybe if they had had to relate to a real-life human the outcome would be different. After all, we know from Chapter 3 that humans do not respond to computers as they do to a fellow human or even a dog. But as well as context it is also the case that stable individual traits may influence how we respond to any potential love drug. Let's welcome back to the page a couple of familiar faces, the OXTR gene and rs53576.

The Genetics of Drugs

We learned in Chapter 5 that the OXTR gene, as a consequence of its polymorphic nature, contributes to a considerable proportion of individual difference in the way we experience and behave when we are in love, including our ability to empathise. Individuals who are homozygous for G (GG) at one particularly variable point on the gene (single nucleotide polymorphism [SNP] rs 53576) show significantly better empathising skill than those who are either homozygous or heterozygous for A (AA or AG) at this point. But it is becoming apparent that whether or not you carry a G allele at rs53576 also has a significant impact on whether or not a potential love drug, such as MDMA, has any effect on you at all, and this is because of the role for oxytocin in the influence that MDMA has upon our brain and our behaviour.

In their study exploring the role for OXTR variation in the influence of MDMA on subjective experience, behavioural neuroscientist Anya Bershad and her team administered 0.75 mg/kg or 1.5 mg/kg of MDMA or a placebo to thirty-nine male and twenty-nine female users of ecstasy. They were told that the research was to explore individual differences in drug responses and that they might receive a stimulant (e.g. ecstasy or an amphetamine), a sedative (e.g. Valium), a hallucinogen (e.g. LSD), a cannabinoid (e.g. marijuana) or a placebo. Sounds like quite a party. Before administration, participants gave a baseline measure of subjective feelings (e.g. anxious, restless, sociable, playful) and blood pressure and heart rate were obtained. They were then given MDMA or the placebo and then on the hour and half-hour they repeated the subjective and cardiovascular measures for four hours. Obviously before all of this had happened they had provided a generous sample of saliva to enable their version of rs53576 to be genotyped.

Results showed that twenty-eight participants were GG, thirty were AG and ten were AA at rs53576 and that overall both doses of MDMA had increased ratings of sociability, euphoria, anxiety and dizziness and measures of cardiovascular function. However, when split by genotype it was only those who carried the G allele who reported feelings of increased sociability at the 1.5 mg/kg MDMA dose. The AA individuals showed no increase in sociability and less of an increase in euphoria as compared to baseline. Genotype had no impact on increases in cardiovascular measures, dizziness or anxiety, which are common side effects of MDMA consumption. Bershad and her team concluded that oxytocin does play a role in the effects of MDMA on individuals and that those who are AA experience no impact on sociability. This has pretty serious consequences not only for those who wish to legalise MDMA for use in psychotherapy, but also for its potential as a universal elixir of love.

An Ethical Conundrum

It is the case that MDMA has shown some efficacy in the treatment of PTSD and has the potential to be used in couples therapy. In their book *Love is the Drug*, ethicists Brian Earp and Julian Savulescu build a convincing case for using MDMA to help couples to increase their trust and openness during counselling, with the long-term outcome that more relationships will stand the test of time. They argue that we are comfortable to use interventions to deal with the psychological mind states which might threaten our closest relationships, for example depression or anxiety, so why would we not turn to biological – drug – interventions if they are available? After all, what is love but a set of neurochemicals – endogenous addictive drugs – which motivate us to start and then stay in our relationships? Wouldn't we simply be supplementing these as we might supplement serotonin with an antidepressant such as an SSRI? And because we know that healthy relationships have such a positive impact on our mental and physical health and our life satisfaction, aren't we duty bound to prescribe such drugs if they are available to improve someone's life experience? Maybe, and Brian and Julian certainly present a picture of a love-drug utopia, albeit a tightly controlled one, where we can all only benefit from the availability of these drugs. But one reason to question their use might be that we do not exist in a perfect world where the rules that are made around the usage of medical drugs are adhered to by everyone, and not everyone uses drugs for positive ends. Releasing love drugs onto the market has the potential to open the gates to an ethical nightmare. And this is no more clearly illustrated than in the search for a cure for 'love sickness'.

The Eternal Sunshine of the Spotless Mind

Many of you probably enjoyed, as I did, the 2004 film *The Eternal Sunshine of the Spotless Mind*. It's an interesting mix between sci-fi and romantic fiction and follows the journey of a couple, Clementine and Joel, who, after a nasty break-up, undergo a medical procedure which erases any memories of each other from their minds. It was a powerful and thought-provoking film which generated lots of discussion about relationships and whether their pain is a necessary part of their experience. However, back in the real world, the question is whether it would ever be possible to create a drug which allowed us to extricate ourselves painlessly from doomed relationships or one-sided infatuations. Or could we even clear our memory completely of the hurt that relationships can inflict?

When I ask my interviewees whether there are any negative sides to love, more regularly than not they will allude to the pain of breaking up. I imagine that many of us would like to alleviate this pain when we experience it. And there is already a drug in common use, the side effects of which might help us to achieve this goal. These are the SSRIs which are commonly used to treat depression. One of the side effects of SSRIs is a drop in libido and a dampening down of emotions, including recognising your own emotions and caring about the emotions of your partner, which makes them a potential candidate for reducing feelings of love and allowing those in damaging or unhealthy relationships to break the bond and leave. Alternatively, they might have a role in reducing the obsessiveness which can lead to the extreme jealousy which can blight some people's relationships throughout their lives. Brian and Julian certainly think this is a possibility, and I agree with them that where they might assist someone whose mental or physical health might be at risk they could be beneficial. But here's a little story about SSRIs which Brian and Julian relate in their book and which makes the alarm bells ring for me because it provides a snapshot of what might happen if we unleash love drugs on the world and they fall into the wrong hands.

Brian and Julian recount an incident of SSRI prescription which sits well outside what most people would find acceptable use. The Orthodox Jewish faith, like many religious faiths, has rules about love, marriage and acceptable forms of sexuality. In an attempt to ensure that devout followers of the faith adhere to these rules, some rabbis and Orthodox Jewish marriage counsellors in Israel have taken to prescribing SSRIs, not for their intended purpose of supporting those who experience depression, but to suppress the sexual urges of young male yeshiva students to ensure that they comply with the laws regarding love and sexuality. And those laws proscribe no sex before marriage and definitely no homosexuality.

You see, we can put in place as many rules as we like about how these drugs should be used only in controlled psychotherapeutic settings, with ongoing support for clients, but unfortunately we do not have the levels of control which Brian Earp and Julian Savulescu believe are possible. It is very likely that should we open this particular stable door and openly advertise the 'love' potential of these drugs, then at the very least repressive regimes will gain a weapon in their fight against what they perceive to be immoral forms of love. Remember that seventy-two countries still deem homosexuality to be illegal. It is not a massive leap of imagination to envisage the use of SSRIs to 'cure' people of this 'affliction'. We only have to look at the existence of conversion therapy to see that this is a distinct possibility.

Beyond this, even within our own relationships, taking these drugs threatens the possibility of opening oneself up to risk or being on the wrong end of a power imbalance. The fact that MDMA makes people less able to recognise negative emotions in others opens users up to risk as they may remain in relationships which are, in fact, not good for them. While Brian and Julian argue for their use only in established relationships, again the lack of control we have once the horse has bolted means that the possibility exists that a whole relationship could be predicated on their use from the start, and this allows for the possibility that love can be turned on and off on a whim. Here's a little anecdote from an interview in *Wired* to illustrate this:

'I met her in the pub,' he says. 'She was nice, attractive, but I wasn't really into her. [Then] the E kicked in and of course, that changed. At the time, going to school, I would only see her on a Saturday night and I was always on E. Of course, love blossomed, and I kind of went with it: Saturday night loved up with her and Sunday afternoon feeling like hugs in the park. The relationship went on for

two years.

'It was great – but not real. Sadly, it was real for her, and when we moved into a flat together it lasted a few months. I really hurt her without meaning to. Without the drug, there was no love on my side . . . She was a great person: attractive and intelligent. The truth, though, was the MDMA created this false sense of love.' **Wired**, 27 August 2017

As an individual you arguably have the autonomy, at least in the 'free' West, to take whatever you want to influence your own behaviour and experience, but the ethical problem arises when the choices you make impact other people. James's girlfriend invested her dreams and years of her life in a relationship which was ultimately built on a very flimsy scaffold of chemicals, the absence of which made her relationship come crashing down. Now in James's case he was unaware of the side effects, but where this knowledge is known, the possibility exists that a partner can be coerced to take a love drug to blind them to the negative behaviour displayed by their other half, or if one person alone takes the drug they potentially have the power to turn love on and off like a light switch. Further, what happens if we accept these drugs for adult therapy and their use bleeds into therapy between parent and child? There is already evidence from rodent studies that repeated administration of oxytocin to young males leads to deficits in the ability to form functioning relationships with parents and mating partners as an adult.

Humans crave control, and arguably the most uncontrollable aspect of our intimate lives is love. But the issues with trying to use our vast knowledge to control love are twofold. In the first instance, this knowledge is incomplete. At present, we simply do not have the scientific knowledge about the long-term effects of these drugs and the impact that individual difference and context have upon their efficacy for us to confidently step forward and say love drugs are something that will benefit our lives. Secondly, we do not live in a utopian world where everyone respects boundaries and people are not repressed or coerced. If we open the floodgates to their use, then we have to accept that we cannot then push the water back if those boundaries are breached or these drugs become yet another way to wield power over those who we perceive to be weaker than ourselves, whether this be a particular 'immoral' population or our partner. The use of love drugs is a massive ethical dilemma and I do not have the answers. The concern remains that the financial reward due to a company of producing these drugs may blind them to these ethical concerns. As a consequence, it is up to us to all take the time to consider whether the potential benefits to our health and happiness really outweigh what might be considerable costs. Do we want to control love so much that we would be willing to accept the risks? Over to you. Would you take a love drug?

This chapter has been about the darker side of love – jealousy, Dark Triad personalities, abusive relationships, despotic but charismatic leaders and the question of love drugs – but to close my exploration of what love is, I want to return to its more positive side and consider love not only as a feeling but as a call to arms. A motivator to achieve great heights and make great changes. Love not as control but as a source of motivation.

CHAPTER TEN

MOTIVATION

● ● ● ●

Brace yourself.

Emotion consists of neural circuits (that are at least partially dedicated), response systems, and a feeling state/process that motivates and organizes cognition and action. Emotion also provides information to the person experiencing it, and may include antecedent cognitive appraisals and ongoing cognition including an interpretation of its feeling state, expressions or social-communicative signals, and may motivate approach or avoidant behavior, exercise control/regulation of responses, and be social or relational in nature. **Carroll Izard** (Izard 2010:363–70).

Wow. That is quite a paragraph. Like love, the term 'emotion' seems to defy definition and is the subject of endless debate and academic articles. But luckily for us, we are not going to head down a definition rabbit hole because love is not an emotion. Is that a bit of a shock? I am not alone in having described love as an emotion in the past (I'm with Darwin on this, so don't feel quite as bad . . .) and I have spent a long time trying not to resort to describing love as such in this book. Describing love as an emotion appears to be a handy catch-all term for a phenomenon which is hard to reduce to one descriptor. However, even those who study emotions, and who are in agreement that it definitely isn't one of our primary emotions (these include disgust, fear and happiness), shy away from placing it among the secondary emotions, e.g. nostalgia or jealousy, such is its complexity and lifelong influence. No. Rather, love is like hunger, thirst and fatigue: a motivation or drive to ensure we seek out the resources which are fundamental to our survival. And as we know from Chapter 1, love is very much essential for our survival.

In this chapter, I am going to explore love not as an emotion but as a motivator. We will look at the scientific explanation as to why we might struggle to confine love to the box labelled 'emotion'. We will learn that whether or not we are one of life's motivated individuals or find it a struggle to get off our sofa, out of our onesie and embrace the day is in part down to our genes and, in particular, the genes associated with our old friend dopamine. We will hear stories about those who have been motivated by love to achieve great feats of endurance, of exploration and even those who have been 'rescued' from lives of crime, and in all likelihood death, so powerful is their love for their child. And we will learn how art, the creation of which many of us turn to as a way to express or understand our love, might in fact be a terribly clever evolutionary trick to advertise our creative, intelligent and downright sexy brains to our potential romantic lovers. It's time to let love out of its box.

Let's Get Vigorous

We learned in Chapter 2 that the first stages of all types of love – romantic, parental, platonic – are underpinned by two key neurochemicals: oxytocin and dopamine. While oxytocin opens you up to making new relationships, it is dopamine, the hormone of vigour, or more usefully here, motivation, which makes sure you receive the kick up the bottom required to actually make a move across the bar, towards the baby or into the playground game, whatever the context of the relationship formation might be. And while oxytocin and dopamine fade into the background somewhat when beta-endorphin kicks in, they are always there, together.

The role for dopamine in motivation has been confirmed both in maternal and reproductive 'love' in rats and monogamous voles. Chemicals which increase the amount of dopamine in circulation increase maternal behaviour towards pups such as licking and retrieving, while chemicals which block its effects reduce these behaviours. In humans, the pioneering work by Dr Helen Fisher and her team has confirmed the role for dopamine in love and its absence underpins apathy. Apathy is defined as a reduction in voluntary, goal-directed behaviour; a lack of motivation. Our apathy can be behavioural – a reluctance to carry out a physical activity; cognitive – a reluctance to apply our minds; and emotional – a reluctance or inability to engage with emotions. While we all experience apathy from time to time – the completion of some of my more demanding research projects has certainly been the victim of cognitive and physical apathy at times, usually when the sun is shining and a gin and tonic is calling – it is a recognised symptom of a range of disorders, including Parkinson's disease, dementia, stroke, depression and schizophrenia. The fact that it is implicated in Parkinson's, which is characterised by the degeneration of dopaminergic neurons in the brain, is strong evidence of dopamine's crucial role in motivation. But apathy's presence in sufferers of Alzheimer's, who exhibit lowered dopamine transporter levels, and schizophrenia, where transmission of dopamine in the prefrontal cortex is negatively impacted, re-inforce these conclusions. Dopamine, a vital neurochemical in the rewarding experience of love, is a motivator and its absence can leave us listless and unfocused.

Add to this the close overlap between the neural fingerprint of love and the dopaminergic motivational circuits of the brain and weight is added to the argument that love is not an emotion but a drive. In her 2014 paper focusing on the role of dopamine in motivation the perfectly named neuroscientist Dr Tiffany Love detailed the areas of the brain which comprise the dopaminergic motivational circuits: the VTA, nucleus accumbens, hippocampus, amygdala, ventral pallidum and prefrontal cortex. Sound familiar? This is also, largely, the neurocircuitry of love. In particular, the nucleus accumbens, which is crammed with dopamine receptors and a focus of intense activity at the start of a relationship, is especially key in motivation, integrating information on the goal or target from the prefrontal cortex, on the nature of the environment from the hippocampus and on the emotional significance of the occasion from the amygdala, and then prompting motor action – that's physical action – via its connections to the ventral pallidum and midbrain, the site of motor control. The fingerprint of love *is* the fingerprint of motivation.

Love: One of Life's Essentials

Love is one of the core needs of a human being. Everybody has a little void within themselves for love. It is something that can be irrational. You cannot always justify why you love somebody or something. When I find something I love, it completes, like a puzzle. **Margo**

In his ground-breaking 2019 paper 'Love is a physiological motivation (like hunger, thirst, sleep or sex)', Spanish psychologist Enrique Burunat argues that for years we have misclassified love, placing it in the box labelled 'emotion' alongside fear, anger, disgust, etc. when, in fact, it should join the other survival-focused motivations which he helpfully lists in his title. He argues that the close parallels between love and addiction – a powerful and often uncontrollable motivation towards a particular object or activity – make it clear that love is a driver which, yes, encompasses a range of emotions covering the spectrum from lust to fear, anger to happiness, but is much more than a feeling. Indeed, it is the longevity of love which strongly argues for its unsuitability as an emotion. Emotions are experienced for relatively short periods of time. They occur in reaction to a stimulus and have evolved to cope with the here and now; fear makes us run, disgust makes us avoid the mouldy food. This does not mean we 'feel' love all the time, but if we were to ask a member of a securely attached, long-term couple whether they were 'in love' I would argue that, regardless of how they felt at that moment, they would answer 'yes'. Further, the response to an emotion is behavioural and physiological. As such, lust, a primary emotion, quickens the heart rate, dilates the pupils and has the potential to lead to sex, but it is an entirely unconscious process. In contrast, love is made up of emotions but also of behaviours, physiological processes *and* thoughts which engage the conscious areas of the brain.

So what is a motivation? Enrique argues that a motivation is a mechanism which ensures that all our survival-based needs are met and our body and brain carry on ticking over nicely. Think of the motivation to find food or water. This does not mean the feeling associated with the mechanism – say, hunger or thirst – is present all the time but it is a mechanism which stays with us for the life course and the 'feeling' which drives the motivation is triggered when the body moves away from being in balance. When our cells lack the energy from food they need to operate and keep us alive, we feel hungry, motivating us to find food. In the same way, love is a constant in our life but we only yearn for it when it is absent, thus motivating us to head out and find it. We are not permanently thirsty, hungry, tired or lovesick. And as a lack of food, water or sleep in childhood impacts our development, so a lack of love has major consequences for our health and wellbeing throughout life, as we know from Chapters 1 and 3. The same cannot be said for a lack of fear, disgust, anger or even happiness. While the presence or absence of some emotions can be unpleasant or even dangerous, indeed they might threaten our survival, they do not have a direct impact on our development.

The Greatest Motivator of All

The concept of love as a motivation was first alighted upon by the psychologist Abraham Maslow in the mid-1940s. He created a model of human need or motivation based upon a pyramid, the lower levels of which must be met by the individual before higher levels of need can be pursued. Here's a figure of Abraham's idea to help you out:

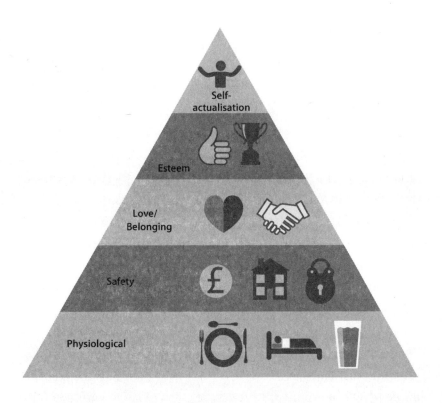

Maslow's pyramid of need

At its base, Maslow's pyramid includes our essential survival-based needs, which are physiological, before moving onwards to the 'psychological' need for relationships in the third level. Maslow conceives of love as a need that must be fulfilled. However, I, along with Enrique Burunat, would argue that love should in fact be placed in level one because, without love, babies are not cared for and our species fails to thrive and we know from Chapter 1 that a lack of love has a direct impact on our physical as well as mental health. Love is, in part, a physiological experience; at the very least, think of all the neurochemicals involved. But despite this, as a model for placing love in the motivation box, rather than defining it as solely an emotion, back in the forties Maslow was bang on the money.

In a recent study the potential for love to be a driver of change was put to the test. After the inevitable paragraph on what love is – emotion, syndrome (one for the philosophers) – the authors, psychologists Agnieszka Hermans-Konopka and Hubert Hermans, present the idea that 'love is a constituent of the self and an instigator of actions in social relationships'. By this they mean that who we love is part of our identity and because we love them we are motivated to take action. Love is a transformative process, an agent of change – a motivator. In particular, love drives people to be the best version of themselves (ignoring the possibility that, as we explored in the last chapter, it can also bring out their darker side) and to reach their full potential.

The people you love bring out your best self. Your happiest self. The person I most enjoy being. When I am with them there is a sort of lifting of 'Oh, not only am I feeling this joy of being with you but I am feeling the joy of being allowed to be this version of me.' There is a self-love that happens when you are with someone else you love that you can only get by being with them. **Judy**

Agnieszka and Hubert recruited 120 Polish university students, aged between twenty-one and twenty-seven, to take part in their study. They wanted to explore the changes in self and motivation caused by fourteen different feelings – note the careful avoidance of the term 'emotion' – including love. These were: self-esteem, love, loneliness, strength, inferiority, anxiety, guilt, safety, tenderness, weakness, joy, anger, internal calm and coldness. Participants were asked to consider each feeling, think of a situation in which they had the feeling and then, using a list of verbs, pick the negative, positive or neutral verbs which reflected how the feeling made them feel about their sense of self and how it might influence their impulses or actions.

Love was amongst the group of feelings, alongside tenderness, inner calm, safety and joy, for both men and women which provoked the fewest negative actions. But of more relevance here, love was the clear winner for both sexes when it came to motivating positive change in someone's life. Agnieszka and Hubert conclude that love is our greatest motivator for positive change and suggest, via their citation of psychologist James Averill, that their findings are unsurprising because love is 'all about reaching out to someone else'. Love *is* all about action towards someone else and Agnieszka and Hubert's work implies that it is quite possibly *the* greatest motivator of all.

The Genetics of Burning Ambition

But the extent to which love acts as a motivator and causes us to 'reach out to someone else' is influenced in part, like so much else, by our genetics. Again, how we behave when we are seeking love or trying to maintain it is in part down to our genes. A recent study, led by Japanese neurologist Shingo Mitaki, explores the role for gene variation in the likelihood that we will experience apathy and, thus, may be less motivated to find love or work to keep it. The key gene she focuses on is one associated with dopamine known as COMT (for science fiends, this stands for catechol-o-methyltransferase). A single nucleotide polymorphism (SNP) causes two different versions of this gene to exist. The result of this is that people who have the Met version have higher levels of dopamine in their prefrontal cortex compared to those who carry the Val version. Shingo and her team wanted to explore whether if a person was Met or Val would affect their likelihood of experiencing apathy. They recruited 963 physically and neurologically healthy volunteers aged between forty-one and eighty-eight years and comprising 513 men and 450 women. They subjected them to the obligatory blood test so that their version of COMT could be assessed and then asked them to complete three self-report measures. These were the Apathy Scale, the Wechsler Adult Intelligence Scale and a measure of depression, itself a cause of apathy.

Their results were clear. Those who carried the Met version had higher dopamine levels, and were significantly less likely to display apathy. Again, we can conclude that dopamine has a clear and central role in apathy. However, while the findings of this study are fascinating – we can attempt to blame our genes next time someone berates us for a lack of ambition or our close relationship with our sofa – it does not conclude that the COMT gene has an influence on any particular form of motivation, say the drive to find love.

However, other studies do hint at a role for variation associated with the COMT gene in the extent to which love might prompt us into action. A recent study by Spanish neuroscientists, led by Leire Erkoreka, has found that Val carriers, who exhibit lower dopamine levels in the prefrontal cortex, are more likely to display avoidant attachment in their romantic relationships. The critical point here is the nature of that attachment: *avoidant*. People who exhibit this style are actively de-motivated from pursuing love. They do not experience the drive or need for love found within secure and anxious attachment styles. For carriers of the Val allele of the COMT gene, love is quite clearly not one of life's great motivators.

Love: A Chance for Re-birth

I think there is something about being a role model. So, I'm thinking about a lot of the things I do or that I let slide. I'm changing the way I do a lot of things and trying to step up my game . . . the thing that is going to make the difference in terms of how he lives his life and who he becomes is actually how I am living my life, day to day. **Will**, dad to Christopher, six months

There is something about being a parent that makes you want to be the best you can be. To want to change, just like Will, to shake things up a bit and embark on this new phase of life with your head held high and an eye on a productive and positive future. I have spent over ten years studying men as they become fathers for the first time and time after time, regardless of who they are, they all voice the same thing: that fatherhood has made them want to up their game. For some, this is giving up smoking, becoming fit or finishing the half-built extension, for others it is more fundamental as they question their work ambitions or take the decision to be a stay-at-home dad. In my book *The Life of Dad: The Making of the Modern Father,* I put this drive for change down to the major shift in identity that accompanies new parenthood, but it may also be due to the huge influx of dopamine that is triggered by the love we have for our children. The changes in psychology and neurochemistry prompted by new parenthood and the love you have for your baby combine to make new parents self-motivating machines. Indeed, in some cases the motivation that new parenthood brings is a life saver for both you *and* your baby.

In the early 2000s American midwife Dr Jenny Foster embarked on a study of the experiences of young Puerto Rican fathers in the United States. All of the dads she followed were the fathers of babies born to mothers who were teenagers at the time of the birth. All of the dads were between the ages of fourteen and twenty-four and reported an income of well below the poverty line. Twelve of the thirty who took part freely admitted they had either been in jail or were involved in illegal activities including the drugs trade. Jenny asked the men a series of questions about their experiences as dads. What was day-to-day life like? Was the pregnancy planned? What did they think was the most important aspect of being a dad and where did their model of fathering come from? She analysed the interview transcripts to look for any universal themes and two stood out: planned pregnancy and, the most interesting for us, the motivating power of being a dad. Dad after dad talked about how having kids made him want to be more responsible and be the positive role model his child needed. In particular, dads were concerned about how their children would view them when they were older, should they continue on their current illegal path. One dad talked about how he had to make the conscious choice to move away from gang life so his kid did not follow his example later in life:

One day I could be walking down the street and I owe somebody money . . . and my son is with me . . . a lot of people don't have compassion. If I kill you with your kid, whatever, pull out a gun in front of your kid, you whole mind is like, 'What?' You don't want to see your son live without you . . . You need a father figure and you can't have nobody else giving him a father figure that you was born to. The only person who can do that is your father. I gotta show my kid the example of that. Yeah, Daddy did this. But he tried to change everything so he can help you. So he don't have to do what I did. **Anonymous gang member** (Foster, 2004: 122)

While these dads had the love of their child to motivate them to make the difficult move of extricating themselves from gang life, history and modern culture are littered with stories of great feats carried out in the name of love. One of the greatest models of romantic love, the medieval chivalric code, is all about love as action: rescuing princesses from the dragons and towers which apparently littered the landscape, and participating in the joust to capture the lady's hand. Starting wars in the name of love also seems to be a thing. In Greek mythology, the Spartan King Menelaus started the Trojan War, which endured for ten years, just to recapture his wife, Helen, from the Trojans. While in the eighteenth century Isabel Godin des Odanais waited twenty years and survived being lost alone in the Amazon rainforest when all other members of her expedition died, just to be reunited with her husband, French cartographer Jean Godin, who had got lost, ironically, while on a map-plotting expedition.

If we bring our analysis of love as a motivator right up to date, the power of love to enable us to cross what appear to be impossible hurdles is evident from recent stories of parents and children, wives and husbands, grandchildren and grandparents who are driven by love to extraordinary feats of physical, emotional and cognitive motivation. Indeed, some of these stories are so powerful and chime such a note of recognition within us that they are deemed to be the stuff that Hollywood blockbusters are made of.

Love on the Silver Screen

She saw my face, after twenty-five years of separation. A mother like her would not have forgotten one of her children's looks. She knew who I was, and I knew who she was. The memory of her face had

been embedded in my mind for such a long time.' **Saroo Brierley,**
Guardian, 24 February 2017.

In 2017 the list for Best Picture at the Oscars read as follows: *La La Land, Arrival, Hell or High Water, Lion, Hidden Figures, Moonlight, Hacksaw Ridge, Manchester by the Sea* and *Fences.* As usual, they were a mix of original screenplays and those based upon real-life figures whose amazing stories merited a turn on the silver screen. However, among them only one was the culmination of a journey which began when the real-life protagonist was five years old and saw him lose his family, be adopted, move halfway around the globe, be reunited with his birth family when he was thirty-one, take part in a documentary and then write a book about his experience which formed the basis of the film's screenplay. And only one film was testament to the motivational power of love.

Lion tells the story of Saroo Brierley. Saroo, then known as Sheru, was born in the early eighties to an impoverished family based in the remote town of Khandwa in Madya Pradesh, India. To help the family survive, he and his two brothers often begged for food and money at the local railway station or managed to obtain work sweeping the floors of railway carriages. One day, when Sheru was five, his elder brother Guddu decided to travel to the city of Burhanpur which was around seventy kilometres away and, despite Guddu's protests, Sheru went with him. When they arrived, Guddu left Sheru on a bench on the platform at Burhanpur while he went to search for work. He did not return and Sheru eventually fell asleep while waiting for him. When he awoke, a train was parked in the station and Sheru, believing Guddu might be on it, boarded the train. He could not find Guddu but, hoping he would come for him, he again fell asleep. When he awoke, the train was moving, eventually arriving in Howrah Station, Kolkata,

1500 kilometres from his home town of Khandwa. Sheru tried to head back home by boarding different trains but ultimately they all returned to Howrah station. He was lost and a very long way from his family. To survive, he scavenged food and begged for money at the train station and on the busy streets of Kolkata. Eventually, after some hair-raising encounters, he was taken to the Indian Society for Sponsorship and Adoption, where he was adopted by an Australian couple, the Brierleys, who took him with them to their home in Tasmania. At the same time as Sheru was waking alone in Kolkata station, his mother was being told that the body of her elder son Guddu had been found – he had been hit by a train – and her younger son Sheru was missing. Despite being poverty-stricken and illiterate, she travelled to Hyderabad, Mumbai, Ajmer, Bhopal and Delhi, a round trip of nearly 4000 kilometres, determined to find her youngest child.

We were tactile, using our hands and faces to express what we felt. The tears spoke for themselves. **Saroo Brierley**, *Guardian*, 24 February 2017.

Twenty-five years later and Saroo, a mispronunciation of Sheru, had been brought up in Tasmania surrounded by the unconditional love of his parents, Sue and John Brierley. He had attended university and worked with his dad in the family business. But the memories of where he came from, and his birth mother, had not gone away. In 2011, spurred on by the map of India that his mum had placed next to his bed as a young child, Saroo began to use Google Earth to try and identify where he came from. He travelled the rail tracks emanating from Howrah Station in Kolkata on the screen, searching for the station where he and Guddu were separated. He remembered that it began with B and ultimately his searching alighted upon Burhanpur Station. He then embarked upon every possible journey from there until he arrived at a station whose landmarks rang a bell in his mind; a fountain where he used to play near the tracks was particularly striking. This was the town of Khandwa and by walking the streets he finally came home. In 2012 Saroo travelled to India to be reunited with his family. He spoke no Hindi and his birth mother no English, but as they touched no words were needed. The power of love spoke for them both.

There was one friend who I was very close to at school. Then she moved away with her parents and we completely lost contact. One day I became really curious and I got a train ticket and just went to the city. I knew where her parents' house was and went up there and knocked on the door and she opened the door. I was quite fortunate. That was the only day she was visiting her mother. From then, we have tried not to break it off any more. **Margo**

Saroo's journey to finding his family was a feat of memory and dogged persistence with the welcome assistance of technology. In contrast, another story emanating from India speaks to the power of sheer physical exertion driven by a determination that others should not suffer the loss of a loved one as the protagonist did. Dashrath Manjhi is known in India as the Mountain Man because his love, and grief, for his wife made him move a mountain. In 1959 Dashrath's wife became critically ill while pregnant with their second child. It was vital that she be taken to a hospital but the route from their remote village, Gehlur in Bihar, to the nearest hospital was blocked by a mountain, the circumventing of which meant a journey of over seventy kilometres. Dasharath had neither the money nor the means to make this long journey sufficiently quickly to save his wife and, as a consequence, she and his unborn baby died. In 1960, spurred on by his love for his wife and the powerful conviction that no one else should experience the deep grief he had endured, he began to hand-carve, using only a hammer and chisel, a route through the mountain to the hospital. The route took him twenty-two years to complete and resulted in a 110-metre-long, 9.1-metre-wide path which cut the travel distance to hospital from seventy kilometres to one. Many thought that he was a lunatic, but his determination, as well as garnering worldwide recognition, his face on stamps and ultimately a state funeral, means that today no one need suffer the pain of loss that Dashrath felt. Indeed, in 2007, after his death, the Indian government replaced his handmade path with a proper road, the presence of which is permanent testament to the loss of a great love and its power to motivate us to achieve the apparently unachievable.

Finally, a story about love with a much happier ending; a forty-five-year marriage which had been predicted in the stars. P. K. Mahanandia grew up a Dalit – an untouchable – in the eastern Indian state of Orissa. Whenever he was upset by the discrimination he and his family experienced at the hands of the higher castes, his mother told him not to worry, his horoscope predicted a happy future for him: marriage to a Taurean woman from a distant land who loved music and owned a jungle. As an adult, P. K. earned his living drawing tourist portraits in a suburb of Delhi. One day, he was asked to draw a portrait by a Swedish woman named Charlotte. As he drew, an easy rapport developed between them and he took the opportunity to ask her for tea. He remembered his mother's prediction and his conversation with her revealed that she ticked all the boxes: a piano-playing, Taurean Swede whose aristocratic family owned a forest. P. K. offered to show her Orissa and as they travelled they fell in love, ultimately marrying in a tribal ceremony at his family home.

Charlotte had travelled by car to India via the popular 'Hippie Trail' and the time came for her to return to Sweden with her friends. She made P. K. promise to join her as soon as he could. But P. K. did not have the luxury of being able to afford a car, so, after a year when their relationship was sustained by nothing more than letter writing, he sold everything he owned, bought a bicycle and on the 22 January 1977 embarked on the 11,000 miles of the hippie trail that stood between him and his love. He travelled for forty-four miles a day and crossed Pakistan, Afghanistan, Iran, Turkey and Europe before arriving in Gothenburg on 28 May 1977. Having initially encountered some resistance to their relationship from her parents, P. K. and Charlotte finally officially married and remain blissfully happy today. P. K. may have had little money and certainly didn't speak the language of those he encountered on his long journey, but he sustained himself by drawing portraits and relying upon the 'universal language of love'. And when people exclaim in wonder at the heroism of his journey he responds: 'I did what I had to do. I was cycling for love but never loved cycling. It's simple.'

The Art of Love

Love is an open secret, the most obvious thing in the world
and the most hidden, with no why to how it keeps its mystery.
Rumi, thirteenth-century Persian poet and scholar

Without love, would we have art? Well, yes we would, the cave paintings of the Upper Palaeolithic in Europe are very likely to have been a source of social power or information exchange rather than an expression of love. And if we trace the origins of creativity further back, the perfectly symmetrical Acheulean handaxes which appear in the archaeological record at around 1.8 million years ago, and whose perfect three-dimensional symmetry is too over-engineered to be of practical use, might mark the point at which craft and art merged, and love is nowhere to be seen. But if we were to word this question more subtly, then it is truly the case that much of our art – our poetry, music, films, paintings and literature – would not exist without love. Think of the number of films, songs, poems and books which are produced by their makers as a result of the need to try and define love, or to capture, even for a moment, the powerful feelings it invokes. Even those among us who might be more artistically challenged might turn to poetry to declare our love for the object of our desire, to attempt to immortalise a significant moment in time or try to untangle our confused feelings. Indeed, we could argue that art and love exist in a feedback relationship. We make art to attempt to understand love, but then use it when we are *in* love to add to our experience. I cannot be the only one who, as a moon-eyed teenager deep in the experience of her first love, listened to love songs on repeat in her bedroom as she fantasised about her boyfriend and wondered in awe at the power of this new feeling.

'Love is not an emotion, it's your very existence.'
Rumi, thirteenth-century Persian poet and scholar

However, this link between love and the making of art is not a phenomenon of the modern age. Long before either Klimt or Rodin tried to capture love in their versions of 'The Kiss', cultures were depicting love through art. Poetry and images from ancient

Egypt show that romantic love was a fully developed concept 1200 years BC. The images of the pharaoh Tutankhamun and his wife Ankhesenamun found in their tomb are renowned for their depiction of devoted love, showing as they do the couple caressing each other, touching each other's faces and exchanging gifts. And while Islam prohibits figurative art, among the best-selling poets in America today is the thirteenth-century Persian poet Jalāl ad-Dīn Muhammad Rūmī, better known as Rumi, whose belief that love is indefinable but sits at the centre of our souls obviously chimes with modern ideas about romantic love.

But the link between art and love may also function at a much more basic level because art, or more broadly, creativity, plays a major role in how we assess the value of a potential mate. In 2000 cognitive psychologist Geoffrey Miller published his book *The Mating Mind: How Sexual Choice Shaped the Evolution of Human Nature*. In it Geoffrey put forward his theory, now widely accepted, that art and creativity evolved not because they have a survival-based role but because they increased the likelihood that the person displaying the creative prowess would be more successful in the mating market. If you remember from Chapter 2, the mating market is like the stock market, we all have a value on our head which is linked to the likelihood we will be reproductively successful and the ability to be creative, Geoffrey argues, significantly increases this value.

But let's start at the beginning. Ever since Darwin observed the many varied finches of the Galapagos, fed his obsession with fossils on several expeditions and came up with the theory of natural selection, the production of art, not simply visual but literary and auditory, has posed a problem. The theory of natural selection states that variation in the natural world is caused by changes in our genes. As a species evolves, those individuals who are better adapted to the environment survive and reproduce, resulting in their genes increasing in frequency in the population and a trait becoming fixed within the species. This is known as 'survival of the fittest'. As a consequence we should be able to explain a species' particular adaptations from the perspective of how they increase the species' chances of survival. However, no one could explain how art increased our chances of our survival. But Geoffrey looked at art a different way. He stated that art doesn't increase our chance of survival but it *does* increase the chance of us being reproductively successful; after all, there is no point surviving if you don't then successfully win a partner and mate. This is a stage of the evolutionary story which Darwin neatly encompassed in his second but lesser-known theory of selection: sexual selection. This theory states that some adaptations don't increase our chance of survival, indeed they might threaten it, but they increase the chance that we can gain a mate and then reproduce. They can do this either by being attractive to the potential mate – they are a form of display – or by equipping us to be better at actively fighting for them against competitors. So the peacock's tail – the evolution of which caused Darwin many fevered nights' sleep – is a result of sexual selection and is used in display by the male to attract a female. It works not because it merely looks pretty but because there is a direct relationship between the size and brightness of a peacock's tail and his genetic health. If as a peacock you can handle your massive tail, which does inhibit your ability to run away from a predator, you are obviously incredibly fit and strong. The female peacock will want that strength, fitness and set of good genes for her offspring. The alternative method is to carry some serious weapons or be of large size. Both of these traits are costly to grow and maintain and may also threaten survival, but they are vital in the often bloody fights for a female which males involve themselves in to ensure their genes do carry on down the generations. Think of the immense size of the male elephant seal or the antlers of the stag. All of these traits have evolved as a result of sexual selection.

Geoffrey argues that art is ultimately no different. The production of art, the possession of creativity, has evolved as a consequence of sexual selection to allow us to advertise the strength of perhaps our most powerful asset: our brains. As such, art is slightly different from the tail of the peacock or the immensity of a silverback's bulk because it is not actually part of our body. This is known as an extended phenotype: a representation of our genetic worth but separate from our body. Humans are not the only animals that produce extended phenotypes as a form of display in mating; the bower bird is well known for the male's production of amazing three-dimensional bowers, the quality of which is a factor in selection by female mates. So Geoffrey argues that art evolved as a consequence of sexual selection to display our genes and, in particular, how those genes have shaped our creativity, intelligence and wit, all of which are evidence for strong cognitive skills and high intelligence. Our brains are the neural equivalent of the peacock's tail.

How does Geoffrey support his argument? Variously by pointing to the tendency for women to request a GSOH (Good Sense of Humour) in dating ads but more powerfully by referencing the immense sexiness of artists and the impact this has upon their reproductive success. I've already referred to the prodigious reproductive abilities of some of our most famous artists in Chapter 1. But these were all wealthy men, which never holds you back in the mating stakes. However, the story of Cyrano de Bergerac also rings so true – the talented but aesthetically challenged poet gets the girl through the exquisiteness of his verse – because to some degree it reflects real experience. There is something about the penniless but passionate artist. Art acts as a concrete example of our creativity, our cognitive flexibility, our intelligence. And intelligence and cognitive flexibility are sexy. As a consequence, our motivation to make great art or write terrible love poetry may not only be a search for the meaning of love but also just another way to attract a mate.

My most powerful love is with my husband and children. I would fly across oceans to rescue and be with them. **Lynn**

This chapter has been about love, not as an emotion but as a motivator. For some, it has been the prompt to great feats of physical and cognitive endurance and it certainly has a role in the production of much of our art. But even for us mere mortals, love motivates us to do perhaps one of the most terrifying things of all: cross the room, engage another person in conversation and, perhaps, alter for ever the course of our lives. What is love? Truly life changing.

EPILOGUE

● ● ● ●

A little while ago, an attendee at one of my talks on the neuro-science of love emailed me. She hadn't had a chance to ask her question at the end of the talk, so had decided to write to me instead. I am really glad she did. She wanted to know, if it were possible for a scientist who studied love to come back from the future having found all the answers, what would be the one question I would ask her? This really made me think. While we know a lot now about how all the different factors come together to influence an individual's experience of love, there is just *so* much to ask. Can animals feel love like we do? Does love for a god confer all the benefits of human relationships? Is there really a difference between how polyamorous and mono-amorous people feel? Why are some people aromantic and what do their brains look like? But then I thought, actually all my questions are really asking the same thing. How does this person feel when they are in love? What is their experience of love like? So, the question I would ask the scientist is this: what is your love like? Because just for one day I would like to live someone else's experience of love. To see what love with their friends, their family, their children, their lover, their god feels like. Because we ultimately have no idea whether what we all say is love is the same thing. Maybe having this opportunity would help me get a little glimpse of what this thing called love actually is and whether what we each feel and experience is even in the same ball park.

In this book I have tried to give you ten solid answers to the why, how, what and who of love. Answers that stand up well as explanations for what this phenomenon is, how it is generated, why we experience it and who we experience it with. But by providing you with ten robust answers I also wanted to make the point that there is no one answer and that love is a highly complex, multifactorial thing. It is awe-inspiring in the true sense of the word. I wanted to excite you about love because I think sometimes we forget how vital and marvellous it is. And I wanted to remind you that love is everywhere, not just with the lover who we are encouraged to dedicate all our spare time to finding. That as humans we are lucky to have this big brain which has enabled us to find love in so many different contexts, and those contexts are still potentially growing as our social world continues to evolve with the advent of AI. In the ten chapters we have explored the evolution of love and encountered the idea – as cold and hard as it might sound – that ultimately love is biological bribery, there to ensure we dedicate time and energy to our survival-critical relationships. That this bribery consists of a set of neurochemicals which motivate us to start and then maintain these relationships. Oxytocin makes us more confident, dopamine rewards us and gives us a push towards action, serotonin gives us obsession, and beta-endorphin? Well, beta-endorphin makes sure we are in our relationships for the long term, be they with lovers, family, children or friends. We have learned that attachment is psychology's attempt to provide an objective measure of love and that it encapsulates those relationships which are the very foundation of our lives, of our health and happiness. That unlike the lesser mammals, we are capable of attaching to many different people, animals and maybe even robots. And that animals feel love too. The devotion of the domestic dog is not to be dismissed any longer as cupboard love, but is the real deal, as all dog owners knew anyway.

We have learned that our love is both private and individual – influenced by our environment, our genes, our gender, sex, ethnicity and age – but also hugely public as our culture shapes what is and is not acceptable in the realm of love, and that for some this means being denied the right to celebrate their love publicly. We have learned that our assumption that romantic love is a one-to-one business is not correct as people become more open about their polyamory or aromanticism, and that monogamy might just be a clever system dreamed up by the powers that be to make sure society remains calm and predictable. That our capacity for love spreads beyond this earthly realm to figures without physical form – our gods – and that these same mechanisms might underpin our fascination with, and our love for, the cult of celebrity. And we have delved into the dark side of love. We have learned that our craving for love, our insatiable desire to have it in our lives, can lead to our exploitation, manipulation and abuse by individuals or groups who wish to control our lives. Finally, we have learned that love cannot be confined to the box labelled 'emotions' but is a motivation, a need as fundamental to life as the food we eat or the air we breathe and the cause of feats of amazing dedication, endurance and strength, all in the name of love.

My ultimate aim with this book is to cause the reader to reconsider the importance of love in their own life, to interrogate their own experience and to reconnect with what lies at the very centre of what it is to be human. To become comfortable with the fact that we will never totally know what love is – a difficult ask for a knowledge-hungry species – but to understand that this unknowability is what makes love the exhilarating, painful but all-encompassing experience it is. To be in awe once more of that most powerful of abilities, our capacity for great love, and to place it once again at the centre of our being. Love is within every bodily fibre and infiltrates every aspect of our external world. Get your relationships right, care for them and all else in life will follow.

So, I cannot give you the complete answer to the question 'What is love?' But by showing you that your experience of love is the result of a myriad of factors, that you can love many people, animals and beings in different ways, and that your ability to love and be loved is fundamental to your wellbeing, I hope I have demonstrated that love does and should sit at the centre of your life. Indeed, maybe the answer to the question 'What is love?' is actually so blindingly obvious that we miss it. What is love? **Everything.**

BIBLIOGRAPHY

• • • •

Chapter 1: Survival

Machin, A. (2018). *The Life of Dad: The Making of the Modern Father*. Simon & Schuster, London.

Dunbar, R. I. M. (2020). 'Structure and Function in Human and Primate Social Networks: Implications for Diffusion, Network Stability and Health'. Proceedings of the Royal Society: Series A.

Holt-Lunstad, J., Smith, T. B. and Layton, J. B. (2010). 'Social Relationships and Mortality Risk: A Meta-analytic Review'. *PLoS Medicine* 7(7): e1000316.

Rodgers, J., Valuev, A. V., Hswen, Y. and Subramanian, S. V. (2019). 'Social capital and physical health: An updated review of the literature for 2007–18'. *Social Science and Medicine* 236: 112360.

Sarkar, D. K., Sengupta, A., Zhang, C., Boyadjieva, N. and Murugan, S. (2012). 'Opiate antagonist prevents mu and delta opiate receptor dimerization to facilitate ability of agonist to control ethanol-altered natural killer cell functions and mammary tumor growth'. *The Journal of Biological Chemistry* 287: 16734–47.

Chapter 2: Addiction

Liebowitz, M. (1983). *The Chemistry of Love*. Little, Brown.

Feldman, R. (2017). 'The Neurobiology of Human Attachments'. *Trends in Cognitive Sciences* 21(2).

Schneiderman, I., Zagoory-Sharon, O, Leckman, J. F. and Feldman, R. (2012). 'Oxytocin during the initial stages of romantic attachment: Relations to couples' interactive reciprocity'. *Psychoneuroendocrinology* 37(8): 1277–85.

Noë, R. and Hammerstein, P. (1995). 'Biological Markets'. *Trends in Ecology and Evolution 10*, 336–9.

Singh, D. (2002). 'Female Mate Value at a Glance: Relationship of Waist-to-Hip Ratio to Health Fecundity and Attractiveness'. *Neuro Endocrinology Letters* 23: 81–91.

Garza, R., Heredia, R. R. and Cieslicka, A. B. (2016). 'Male and Female Perception of Physical Attractiveness: An Eye Movement Study'. *Evolutionary Psychology* 14: 1–16.

Miller, G. (2000). *The Mating Mind: How Sexual Choice Shaped The Evolution of Human Nature*. Vintage.

Machin, A. and Dunbar, R. I. M. (2013). 'Sex and gender as factors in romantic partnerships and best friendships'. *Journal of Relationships Research* 4: e8.

Bartels, A. and Zeki, S. (2000). 'The neural basis of romantic love'. *NeuroReport* 11(117): 3829–34.

Bartels, A. and Zeki, S. (2004). 'The neural correlates of maternal and romantic love'. *NeuroImage* 21: 1155–66.

Atzil, S., Hendler, T., Zagoory-Sharon, O., Winetraub, Y. and Feldman, R. (2012). 'Synchrony and specificity in the maternal and paternal brain: relations to oxytocin and vasopressin'. *Journal of the American Academy of Child and Adolescent Psychiatry* 51(8): 798–811.

Wlodarski, R. and Dunbar, R. I. M. (2016). 'When BOLD is thicker than water: processing social information about kin and friends at different levels of the social network'. *Social Cognitive and Affective Neuroscience*: 195–260.

Machin, A. and Dunbar, R.I.M. (2014). 'The brain opioid theory of social attachment: a review of the evidence' in R. I. M. Dunbar C, Gamble and J. A. J. Gowlett (eds), *Lucy to Language: Benchmark Papers*. Oxford University Press, pp. 181–213.

Nummenmaa, L., Tuominen, L., Dunbar, R., Hirvonen, J., Manninen, S., Helin, S., Machin, A., Hari, R., Jääskeläinen, I. P. and Sams, M. (2016). 'Social touch modulates endogenous μ-opioid system activity in humans'. *NeuroImage* 138: 242–7.

Pearce, E., Wlodarski, R., Machin, A. and Dunbar, R. I. M. (2017). 'Variation in the β-endorphin, oxytocin, and dopamine receptor genes is associated with different dimensions of human sociality'. *PNAS* 114: 5300–5.

Cohen, E. E. A., Ejsmond-Frey, R., Knight, N. and Dunbar, R. I. M. (2009). 'Rowers' high: behavioural synchrony is correlated with elevated pain thresholds'. *Biology Letters*.

Tarr, B., Launay, J. and Dunbar, R. I. M. (2016). 'Silent disco: dancing in synchrony leads to elevated pain thresholds and social closeness'. *Evolution and Human Behavior*.

Tarr, B., Launay, J., Benson, C. and Dunbar, R. I. M. (2017). 'Naltrexone blocks endorphin released when dancing in synchrony'. *Adaptive Human Behavior and Physiology* 3: 214–54.

Pearce, E. (2016). 'Participants' perspectives on the social bonding and well-being effects of creative arts adult education classes'. *Arts and Health* 9: 42–59.

Ulmer,-Yaniz, A., Avitsur, R., Kanat-Maymon, Y., Schneiderman, I., Zagoory-Sharon, O. and Feldman, R. (2016). *Brain, Behavior and Immunity* 56: 130–9.

Chapter 3: Attachment

Krumwiede, A. (2014). *Attachment Theory According to John Bowlby and Mary Ainsworth*. GRIN Verlag.

J. Cassidy and P. Shaver (eds) (2018). *Handbook of Attachment, Third Edition: Theory, Research, and Clinical Applications*. Guilford Press.

Machin, A. (2018*). The Life of Dad: The Making of the Modern Father*. Simon & Schuster, London.

Paquette, D. and Bigras, M. (2010). 'The risky situation: a procedure for assessing the father-child activation relationship'. *Early Child Development and Care* 180: 33–50.

Abraham, E., Hendler, T., Zagoory-Sharon, O. and Feldman, R. (2016). 'Network integrity of the parental brain in infancy supports the development of children's social competencies'. *Social, Cognitive and Affective Neuroscience* 11: 1707–18.

Bakermans-Kranenburg, M. J., van Ijzendoorn, M. H., Bokhorst, C. L. and Schuengel, C. (2004). 'The importance of shared environment in infant-father attachment: A behavioural genetic study of the attachment Q-Sort'. *Journal of Family Psychology* 18: 545–9.

Fearon P., Shmueli-Goetz Y., Viding E., et al. (2014). 'Genetic and environmental influences on adolescent attachment. *Journal of Child Psychology and Psychiatry* 55(9): 1033–41.

Nummenmaa, L., Manninen, S., Tuominen, L., Hirvonen, J., Kalliokoski, K. K., Nuutila, P., Jääskeläinen, I. P., Hari, R., Dunbar,

R. I. M. and Sams, M. (2015). 'Adult Attachment Style is Associated With Cerebral Mu-Opioid Receptor Availability in Humans'. *Human Brain Mapping* 36:3621–8.

Feldman, R. (2012). 'Bio-behavioral Synchrony: A Model for Integrating Biological and Microsocial Behavioral Processes in the Study of Parenting'. *Parenting: Science and Practice* 12: 154–64.

Levy, J., Goldstein, A. and Feldman, R. (2017). 'Perception of social synchrony induces mother–child gamma coupling in the social brain'. *Social Cognitive and Affective Neuroscience* 12: 1036–46.

Weisman, O., Zagoory-Sharon, O. and Feldman, R. (2012). 'Oxytocin Administration to Parent Enhances Infant Physiological and Behavioral Readiness for Social Engagement'. *Biological Psychiatry* 72: 982–9.

Chapter 4: Underestimated

Dunbar, R. I. M. (2018). 'The Anatomy of Friendship'. *Trends in Cognitive Sciences* 22: 32–51.

Parkinson, C., Kleinbaum, A. M. and Wheatley, T. (2018). 'Similar neural responses predict friendship'. *Nature Communications* 9: 332.

Brumbaugh, C. C. (2017). 'Transferring connections: Friend and sibling attachments' importance in the lives of singles'. *Personal Relationships* 24: 534–49.

Chavez, R. S. and Wagner, D. D. (2019). 'The Neural Representation of Self is Recapitulated in the Brains of Friends: A Round-Robin fMRI Study'. *Journal of Personality and Social Psychology* 118: 407–16.

Ma, X., Zhao, W., Luo, R., Zhou, F., Geng, Y., Gao, Z., Zheng, X., Becker, B. and Kendrick, K. M. (2018). 'Sex- and context-dependent effects of oxytocin on social sharing'. *NeuroImage* 183: 62–72.

Blair, K. L. and Pukall, C. F. (2015). 'Family matters, but sometimes chosen family matters more: Perceived social network influence in the dating decisions of same- and mixed-sex couples'. *The Canadian Journal of Human Sexuality* 24: 257–70.

Zaleski, N. (2013). *Given and Chosen: Talking to Family about Sexuality*. The Illinois Caucus for Adolescent Health.

Muraco, A. (2006). 'Intentional Families: Fictive Kin Ties Between Cross-Gender, Different Sexual Orientation Friends'. *Journal of Marriage and Family* 68: 1313–25.

Solomon, J., Schöberl, I., Gee, N. and Kotrschal, K. (2019). 'Attachment security in companion dogs: adaptation of Ainsworth's strange situation and classification procedures to dogs and their human caregivers'. *Attachment and Human Development* 21: 389–417.

Lem, M., Coe, J. B., Haley, D. B., Stone, E. and O'Grady, W. (2016). The Protective Association between Pet Ownership and Depression among Street-involved Youth: A Cross-sectional Study. *Anthrozoos* 29: 123–36.

Handlin, L., Hydbring-Sandberg, E., Nilsson, A., Ejdebäck, M., Jansson, A. and Uvnäs-Moberg, K. (2011). 'Short-Term Interaction between Dogs and Their Owners: Effects on Oxytocin, Cortisol, Insulin and Heart Rate – An Exploratory Study'. *Anthrozoos* 24: 301–15.

Stoeckel, L. E., Palley, L. S., Gollub, R. L., Niemi, S. M. and Evins, A. E. (2014). 'Patterns of Brain Activation when Mothers View Their Own Child and Dog: An fMRI Study'. *PLOSOne* 9: e107205.

Fullwood, C., Quinn, S., Kaye, L. K. and Redding, C. (2017). 'My virtual friend: A qualitative analysis of the attitudes and experiences of Smartphone users: Implications for Smartphone attachment'. *Computers in Human Behavior* 75: 347–55.

Konok, V., Korcsok, B., Miklósi, A. and Gácsi, M. (2018). 'Should we love robots? The most liked qualities of companion dogs and how they can be implemented in social robots'. *Computers in Human Behavior* 80: 132–42.

Chaminade, T., Zecca, M., Blakemore, S-J., Takanishi, A., Frith, C. D., Micera, S., Dario, P., Rizzolatti, G., Gallese, V. and Umiltà, M. A. (2010). 'Brain Response to a Humanoid Robot in Areas Implicated in the Perception of Human Emotional Gestures'. *PLOSOne* 5: e11577.

Machin, A. (2020). 'Would you want a robot to be your relative's carer?' *Guardian*, 10 September 2020.

Chapter 5: Personal

Gong, P., Fan, H., Liu, J., Yang, X., Zhang, K. and Zhou, X. (2017). 'Revisiting the impact of OXTR rs53576 on empathy: A population-based study and a meta-analysis'. *Psychoneuroendocrinology* 80: 131–6.

Ebbert, A. M., Infurna, F. J., Luthar, S. S., Lemery-Chafant, K. and Corbin, W. R. (2019). 'Examining the link between emotional childhood abuse and social relationships in midlife: The moderating role of the oxytocin receptor gene'. *Child Abuse and Neglect* 98.

Kraaijenvanger, E. J., He, Y., Spencer, H., Smith, A. K., Bos, P. A. and Boks, M. P. M. (2019). 'Epigenetic variability in the human oxytocin receptor (OXTR) gene: A possible pathway from early life experiences to psychopathologies'. *Neuroscience and BioBehavioral Reviews* 96: 127–42.

Ebner, N. C., Lin, T., Muradoglu, M., Weit, D. H., Plasencia, G. M., Lillard, T. S., Pourna-jafi-Nazarloo, H., Coehn, R. A., Carter, C. S. and Connelly, J. J. (2019). 'Associations between the oxytocin receptor gene (OXTR) methylation, plasma oxytocin and attachment across adulthood'. *International Journal of Psychophysiology* 136:22-32.

Feldman, R., Gordon, I., Influs, M., Gutbir, T. and Ebstein, R. (2013). 'Parental oxytocin and early caregiving jointly shape children's oxytocin response and social reciprocity'. *Neuropsychopharmacology* 38:1154–62.

Butovskaya, P. R., Lazebny, O. E., Sukhodolskaya, E. M., Vasiliev, V. A., Dronova, D. A., Fedenok, J. N., Rosa, A., Peletskaya, E. N., Ryskov, A. P. and Butovskaya, M. L. (2016). 'Polymorphisms of two loci at the oxytocin receptor gene in populations of Africa, Asia and South Europe'. *BMC Genetics* 17:17.

Whittle, S., Yücel, M., Yap, M. B. H. and Allen, N. B. (2011). 'Sex differences in the neural correlates of emotion: Evidence from neuroimaging'. *Biological Psychology* 87: 319–33.

Yin, J., Zou, Z., Song, H., Zhang, Z., Yang, B. and Huang, X. (2018). 'Cognition, emotion and reward networks associated with sex differences for romantic appraisals'. *Nature Scientific Reports* 8: 2835.

Brechet, C. (2015). 'Representation of Romantic Love in children's drawings: Age and gender differences'. *Social Development* 24(3): 640–58.

Liu, J., Gong, P. and Zhou, X. (2014). 'The association between romantic relationship status and 5-HT1A gene in young adults'. *Scientific Reports* 4:7049.

Zeki, S. and Romaya, J. P. (2010). 'The brain reaction to viewing faces of opposite and same-sex romantic partners'. *Plos One* 5(12): e15802.

Chapter 6: Public

Jankowiak, W. R. and Fischer, E. F. (1992). 'A cross-cultural perspective on romantic love'. *Ethnology* 31(2): 149–55.

Parkinson, C., Walker, T. T., Memmi, S. and Wheatley, T. (2017). 'Emotions are understood from biological motion across cultures'. *Emotion* 17(3): 459–77.

Chang, L., Wang, Y., Shackelford, T. K. and Buss, D. M. (2011). 'Chinese mate preferences: Cultural evolution and continuity across a quarter of a century'. *Personality and Individual Differences* 50: 678–83.

Karandashev, V. (2017). *Romantic Love in Cultural Contexts*. Springer.

Pilishvili, T. S. and Koyanongo, E. (2016). 'The representation of love among Brazilians, Russians and Central Africans: A comparative analysis'. *Psychology in Russia: State of the Art* 9(1): 84–97.

De Munck, V., Korotayev, A. and McGrevey, J. (2016). 'Romantic love and family organisation: A case for romantic love as a biosocial universal'. *Evolutionary Psychology* 1–13.

Illouz, E. (2012). *Why Love Hurts*. Polity.

Yahya, S., Boag, S., Munshi, A. and Litvak-Hirsch, T. (2016). '"Sadly Not All Love Affairs Are Meant To Be ..." Attitudes Towards Interfaith Relationships in a Conflict Zone'. *Journal of Intercultural Studies* 37(3): 265–85.

Lasser, J. and Tharinger, D. (2003). 'Visibility management in school and beyond: A qualitative study of gay, lesbian, bisexual youth'. *Journal of Adolescence* 26(2): 233–44.

Chapter 7: Exclusive

Burleigh, T. J., Rubel, A. N. and Meegan, D. V. (2017). 'Wanting "the whole loaf": zero-sum thinking about love is associated with prejudice against consensual non-monogamists'. *Psychology and Sexuality* 8: 24–40.

Hutzler, K. T., Giuliano, T. A., Herselman, J. R. and Johnson, S. M. (2016). 'Three's a crowd: public awareness and (mis)perceptions of polyamory'. *Psychology and Sexuality* 7: 69–87.

Klesse, C. (2011). 'Notions of Love in Polyamory: Elements in a Discourse on Multiple Loving'. *Russian Review of Social Research* 3: 4–25.

Wolkomir, M. (2015). 'One But Not the Only: Reconfiguring Intimacy in Multiple Partner Relationships'. *Qualitative Sociology* 38: 417–38.

Rubel, A. N. and Bogaert, A. F. (2015). 'Consensual Nonmonogamy: Psychological Well-Being and Relationship Quality Correlates'. *The Journal of Sex Research* 52: 961–82.

van Anders, S. M., Hamilton, L. D. and Watson, N. V. (2007). 'Multiple partners are associated with higher testosterone in North American men and women'. *Hormones and Behavior* 51: 454–9.

Hamilton, D. L. and Meston, C. M. (2017). 'Differences in Neural Response to Romantic Stimuli in Monogamous and Non-Monogamous Men'. *Archives of Sexual Behavior* 46: 2289–99.

Chapter 8: Sacred

Beck, R. and McDonald, A. (2004). 'Attachment to God: The Attachment to God Inventory, tests of working model correspondence and an exploration of faith group differences'. *Journal of Psychology and Theology* 32: 92–103.

Beauregard, M. and Paquette, V. (2006). 'Neural correlates of a mystical experience in Carmelite nuns'. *Neuroscience Letters* 405: 186–90.

Schjoedt, U., Stødkilde-Jørgensen, H., Geertz, A. W. and Roepstorff, A. (2009). 'Highly religious participants recruit areas of social cognition in personal prayer'. Soc Cogn Affect Neurosci: 4(2): 199-207.

Hackney, C. H. and Sanders, G. S. (2003). 'Religiosity and Mental Health: A Meta-Analysis of Recent Studies'. *Journal for the Scientific Study of Religion* 42: 43–55.

McClintock, C. H., Lau, E. and Miller, L. (2016). 'Phenotypic Dimensions of Spirituality: Implications for Mental Health in China, India, and the United States'. *Frontiers in Psychology* 7: 1600.

Stever, G. S. (2017). 'Evolutionary Theory and Reactions to Mass Media: Understanding Parasocial Attachment'. *Psychology of Popular Media Culture* 6: 95–102.

Bond, B. J. (2018). 'Parasocial Relationships with Media Personae: Why They Matter and How They Differ Among Heterosexual, Lesbian, Gay, and Bisexual Adolescents'. *Media Psychology 21: 457–85.*

Chapter 9: Control

Buss, D. M. (2003). *The Evolution of Desire: Strategies of Human Mating*. Basic Books.

Güçlü, O., Senormanci, Ö., Senormanci, G. and Köktürk, F. (2017). 'Gender differences in romantic jealousy and attachment styles'. *Psychiatry and Clinical Psychopharmacology* 27: 359–65.

Chegeni, R., Pirkalani, R. K. and Dehshiri, G. (2018). 'On love and darkness: The Dark Triad and mate retention behaviors in a non-Western culture'. *Personality and Individual Differences* 122: 43–6.

Nagler, U. K. J., Reiter, K. J., Furtner, M. R. and Rauthmann, J. F. (2014). 'Is there a "dark intelligence"? Emotional intelligence is used by dark personalities to emotionally manipulate others'. *Personality and Individual Differences* 65: 47–52.

Smith, M., Nunley, B. and Martin, E. (2013). 'Intimate Partner Violence and the Meaning of Love'. *Issues in Mental Health Nursing* 34: 395–401.

Singh, S. and Myende, T. (2017). 'Redefining love: Female university students developing resilience to intimate partner violence'. *Agenda: Empowering women for gender equity* 31: 22–33.

van Vugt, M. (2014). 'On faces, gazes, votes, and followers: Evolutionary psychological and social neuroscience approaches to leadership' in J. Decety and Y. Christen (eds), *Research and perspectives in neurosciences: Vol. 21. New frontiers in social neuroscience* (pp. 93–110). Springer International Publishing.

Grabo, A., Spisak, B. and van Vugt, M. (2017). 'Charisma as signal: An evolutionary perspective on charismatic leadership'. *The Leadership Quarterly* 28: 473–485.

Schoedt, U., Sørensen, J., Nielbo, L., Xygalatas, D., Mitkidis, P. and Bulbulia, J. (2013). 'Cognitive resource depletion in religious interactions'. *Religion, Brain and Behavior* 3: 39–86.

Bartz, J. A., Zaki, J., Bolger, N. and Ochsner, K. N. (2011). 'Social effects of oxytocin in humans: context and person matter'. *Trends on Cognitive Sciences* 15: 301–9.

De Dreu, C. K. W., Greer, L. L., Van Kleef, G. A., Shalvi, S. and Handgraaf, M. J. J. (2011). 'Oxytocin promotes human ethnocentrism'. *PNAS* 4: 1262–6.

Bedi, G., Hyman, D. and de Wit, H. (2010). 'Is ecstasy an "empathogen"? Effects of MDMA on prosocial feelings and identification of emotional states in others'. *Biological Psychiatry* 68: 1134–40.

Borissova, A. et al. (2020). 'Acute effects of MDMA on trust, cooperative behaviour, and empathy: A double-blind, placebo-controlled experiment'. *Journal of Psychopharmacology* 35(5): 547–55.

Bershad, A. K., Weafer, J. J., Kirkpatrick, W. G., Wardle, M. C., Miller, M. A. and de Wit, H. (2016). 'Oxytocin receptor gene variation predicts subjective responses to MDMA'. *Social Neuroscience* 11: 592–9.

Earp, B. D. and Savulescu, J. (2020). *Love is the Drug: The Chemical Future of our Relationships*. Manchester University Press.

Chapter 10: Motivation

Love, T. M. (2014). 'Oxytocin, motivation and the role of dopamine'. *Pharmacology, Biochemistry and Behavior*: 49-60.

Chong, T. (2018). 'Updating the role of dopamine in human motivation and apathy'. *Current Opinion in Behavioral Sciences* 22: 35–41.

Fisher, H., Aron, A. and Brown, L. L. (2005). 'Romantic love: An fMRI study of a neural mechanism for mate choice'. *The Journal of Comparative Neurology* 493: 58–62.

Burunat, E. (2019). 'Love is a physiological motivation (like hunger, thirst, sleep or sex)'. *Medical Hypotheses* 129.

Maslow, A. H. (1943). 'A theory of human motivation'. *Psychological Review* 50: 370–96.

Hermans-Konopka, A., & Hermans, H. J. M. (2010). *The dynamic features of love: Changes in self and motivation.* In J. D. Raskin, S. K. Bridges, & R. A. Neimeyer (eds.), *Studies in meaning 4: Constructivist perspectives on theory, practice, and social justice* (p. 93–123). Pace University Press.

Mitaki, S., Isomura, M., Maniwa, K., Yamasaki, M., Nagai, A., Nabika, T. and Yamaguchi, S. (2013). 'Apathy is associated with a single-nucleotide polymorphism in dopamine-related gene'. *Neuroscience Letters* 549:87-91.

Erkoreka, L., Mercedes, Z., Macias, I., Angel Gonzalez-Torres, M. (2018). 'The COMT Val158Met polymorphism exerts a common influence on avoidant attachment and inhibited personality, with a pattern of positive heterosis'. *Psychiatry Research, 262,* 345-7.

Foster, J. (2004). 'Fatherhood and the Meaning of Children: An Ethnographic Study Among Puerto Rican Partners of Adolescent Mothers'. *Journal of Midwifery and Women's Health* 49: 118-25.

Miller, G. (2001). *The Mating Mind: How sexual choice shaped the evolution of human nature.* Vintage.

ACKNOWLEDGEMENTS

• • • •

In the first instance I would like to thank my agent Sally Holloway and the team at Felicity Bryan Associates. Sally has always offered a listening ear and a calming voice when it has been most needed. Her eye for what will be popular and her feedback on early drafts has always made sure I keep heading in the right direction. To my editor, Maddy Price, and to Clarissa Sutherland, Ellen Turner, Britt Sankey and the rest of the team at Orion, I can say that without exception working with them on this book has been a joy and pleasure. And I must offer Maddy extra thanks for suggesting I take some time to interview dog owners and their adorable hounds. Without her I would not have had my best Zoom experience by far during lockdown. I need to thank my (ex-)boss Professor Robin Dunbar. He took a chance on me thirteen years ago and has supported my career ever since. He has kindly read drafts of this book and answered my questions when my memory for a particular paper or experiment has failed me. He is my example and mentor and I will be forever grateful to him. To my colleagues at SENRG Oxford, in particular Ellie Pearce and Rafael Wlodorski. Exploring the science of love with them was an exciting and creative time despite the spit. To my best friends: the two Fis. To my oldest and dearest friend Ffiona. We have been parted by the Atlantic and Covid but our decades-long love for each other endures. And to Fi, my much closer neighbour who has provided support, gossip, an empathetic ear and home-made cake when it has been needed. I can't wait for a big night out to celebrate publication. To my dogs, Bear, Sam and Scruffy, who as my office colleagues can always be relied upon to squeak a toy, snore or woof at the most inopportune moment, but who are all the proof I need that what exists between dog and owner is love. To my family. To my stepdaughter Lydia who was the first to teach me about parental love and to my parents, Liz and Norman. I can honestly say that none of this would have been possible without them, from funding my education, to reading my many, many draft papers to being the best hands-on grandparents to my daughters. Being a mum and an academic can be a tough gig and I can say without any doubt that without them I would not have managed to balance the love I have for my career with my absolute determination to be a hands-on mum. And finally, to the greatest loves of my life: Julian, Hebe and Kitty. They are my world and my whole reason for being. Thank you.